ROC Curves for Continuous Data

MONOGRAPHS ON STATISTICS AND APPLIED PROBABILITY

General Editors

J. Fan, V. Isham, N. Keiding, T. Louis, R. L. Smith, and H. Tong

1 Stochastic Population Models in Ecology and Epidemiology *M.S. Barlett* (1960)
2 Queues *D.R. Cox and W.L. Smith* (1961)
3 Monte Carlo Methods *J.M. Hammersley and D.C. Handscomb* (1964)
4 The Statistical Analysis of Series of Events *D.R. Cox and P.A.W. Lewis* (1966)
5 Population Genetics *W.J. Ewens* (1969)
6 Probability, Statistics and Time *M.S. Barlett* (1975)
7 Statistical Inference *S.D. Silvey* (1975)
8 The Analysis of Contingency Tables *B.S. Everitt* (1977)
9 Multivariate Analysis in Behavioural Research *A.E. Maxwell* (1977)
10 Stochastic Abundance Models *S. Engen* (1978)
11 Some Basic Theory for Statistical Inference *E.J.G. Pitman* (1979)
12 Point Processes *D.R. Cox and V. Isham* (1980)
13 Identification of Outliers *D.M. Hawkins* (1980)
14 Optimal Design *S.D. Silvey* (1980)
15 Finite Mixture Distributions *B.S. Everitt and D.J. Hand* (1981)
16 Classification *A.D. Gordon* (1981)
17 Distribution-Free Statistical Methods, 2nd edition *J.S. Maritz* (1995)
18 Residuals and Influence in Regression *R.D. Cook and S. Weisberg* (1982)
19 Applications of Queueing Theory, 2nd edition *G.F. Newell* (1982)
20 Risk Theory, 3rd edition *R.E. Beard, T. Pentikäinen and E. Pesonen* (1984)
21 Analysis of Survival Data *D.R. Cox and D. Oakes* (1984)
22 An Introduction to Latent Variable Models *B.S. Everitt* (1984)
23 Bandit Problems *D.A. Berry and B. Fristedt* (1985)
24 Stochastic Modelling and Control *M.H.A. Davis and R. Vinter* (1985)
25 The Statistical Analysis of Composition Data *J. Aitchison* (1986)
26 Density Estimation for Statistics and Data Analysis *B.W. Silverman* (1986)
27 Regression Analysis with Applications *G.B. Wetherill* (1986)
28 Sequential Methods in Statistics, 3rd edition
G.B. Wetherill and K.D. Glazebrook (1986)
29 Tensor Methods in Statistics *P. McCullagh* (1987)
30 Transformation and Weighting in Regression
R.J. Carroll and D. Ruppert (1988)
31 Asymptotic Techniques for Use in Statistics
O.E. Bandorff-Nielsen and D.R. Cox (1989)
32 Analysis of Binary Data, 2nd edition *D.R. Cox and E.J. Snell* (1989)
33 Analysis of Infectious Disease Data *N.G. Becker* (1989)
34 Design and Analysis of Cross-Over Trials *B. Jones and M.G. Kenward* (1989)
35 Empirical Bayes Methods, 2nd edition *J.S. Maritz and T. Lwin* (1989)
36 Symmetric Multivariate and Related Distributions
K.T. Fang, S. Kotz and K.W. Ng (1990)
37 Generalized Linear Models, 2nd edition *P. McCullagh and J.A. Nelder* (1989)
38 Cyclic and Computer Generated Designs, 2nd edition
J.A. John and E.R. Williams (1995)
39 Analog Estimation Methods in Econometrics *C.F. Manski* (1988)
40 Subset Selection in Regression *A.J. Miller* (1990)
41 Analysis of Repeated Measures *M.J. Crowder and D.J. Hand* (1990)
42 Statistical Reasoning with Imprecise Probabilities *P. Walley* (1991)
43 Generalized Additive Models *T.J. Hastie and R.J. Tibshirani* (1990)

44 Inspection Errors for Attributes in Quality Control
N.L. Johnson, S. Kotz and X. Wu (1991)

45 The Analysis of Contingency Tables, 2nd edition *B.S. Everitt* (1992)

46 The Analysis of Quantal Response Data *B.J.T. Morgan* (1992)

47 Longitudinal Data with Serial Correlation—A State-Space Approach
R.H. Jones (1993)

48 Differential Geometry and Statistics *M.K. Murray and J.W. Rice* (1993)

49 Markov Models and Optimization *M.H.A. Davis* (1993)

50 Networks and Chaos—Statistical and Probabilistic Aspects
O.E. Barndorff-Nielsen, J.L. Jensen and W.S. Kendall (1993)

51 Number-Theoretic Methods in Statistics *K.-T. Fang and Y. Wang* (1994)

52 Inference and Asymptotics *O.E. Barndorff-Nielsen and D.R. Cox* (1994)

53 Practical Risk Theory for Actuaries
C.D. Daykin, T. Pentikäinen and M. Pesonen (1994)

54 Biplots *J.C. Gower and D.J. Hand* (1996)

55 Predictive Inference—An Introduction *S. Geisser* (1993)

56 Model-Free Curve Estimation *M.E. Tarter and M.D. Lock* (1993)

57 An Introduction to the Bootstrap *B. Efron and R.J. Tibshirani* (1993)

58 Nonparametric Regression and Generalized Linear Models
P.J. Green and B.W. Silverman (1994)

59 Multidimensional Scaling *T.F. Cox and M.A.A. Cox* (1994)

60 Kernel Smoothing *M.P. Wand and M.C. Jones* (1995)

61 Statistics for Long Memory Processes *J. Beran* (1995)

62 Nonlinear Models for Repeated Measurement Data
M. Davidian and D.M. Giltinan (1995)

63 Measurement Error in Nonlinear Models
R.J. Carroll, D. Rupert and L.A. Stefanski (1995)

64 Analyzing and Modeling Rank Data *J.J. Marden* (1995)

65 Time Series Models—In Econometrics, Finance and Other Fields
D.R. Cox, D.V. Hinkley and O.E. Barndorff-Nielsen (1996)

66 Local Polynomial Modeling and its Applications *J. Fan and I. Gijbels* (1996)

67 Multivariate Dependencies—Models, Analysis and Interpretation
D.R. Cox and N. Wermuth (1996)

68 Statistical Inference—Based on the Likelihood *A. Azzalini* (1996)

69 Bayes and Empirical Bayes Methods for Data Analysis
B.P. Carlin and T.A Louis (1996)

70 Hidden Markov and Other Models for Discrete-Valued Time Series
I.L. MacDonald and W. Zucchini (1997)

71 Statistical Evidence—A Likelihood Paradigm *R. Royall* (1997)

72 Analysis of Incomplete Multivariate Data *J.L. Schafer* (1997)

73 Multivariate Models and Dependence Concepts *H. Joe* (1997)

74 Theory of Sample Surveys *M.E. Thompson* (1997)

75 Retrial Queues *G. Falin and J.G.C. Templeton* (1997)

76 Theory of Dispersion Models *B. Jørgensen* (1997)

77 Mixed Poisson Processes *J. Grandell* (1997)

78 Variance Components Estimation—Mixed Models, Methodologies and Applications *P.S.R.S. Rao* (1997)

79 Bayesian Methods for Finite Population Sampling
G. Meeden and M. Ghosh (1997)

80 Stochastic Geometry—Likelihood and computation
O.E. Barndorff-Nielsen, W.S. Kendall and M.N.M. van Lieshout (1998)

81 Computer-Assisted Analysis of Mixtures and Applications—
Meta-analysis, Disease Mapping and Others *D. Böhning* (1999)

82 Classification, 2nd edition *A.D. Gordon* (1999)

83 Semimartingales and their Statistical Inference *B.L.S. Prakasa Rao* (1999)
84 Statistical Aspects of BSE and vCJD—Models for Epidemics
   C.A. Donnelly and N.M. Ferguson (1999)
85 Set-Indexed Martingales *G. Ivanoff and E. Merzbach* (2000)
86 The Theory of the Design of Experiments *D.R. Cox and N. Reid* (2000)
87 Complex Stochastic Systems
   O.E. Barndorff-Nielsen, D.R. Cox and C. Klüppelberg (2001)
88 Multidimensional Scaling, 2nd edition *T.F. Cox and M.A.A. Cox* (2001)
89 Algebraic Statistics—Computational Commutative Algebra in Statistics
   G. Pistone, E. Riccomagno and H.P. Wynn (2001)
90 Analysis of Time Series Structure—SSA and Related Techniques
   N. Golyandina, V. Nekrutkin and A.A. Zhigljavsky (2001)
91 Subjective Probability Models for Lifetimes
   Fabio Spizzichino (2001)
92 Empirical Likelihood *Art B. Owen* (2001)
93 Statistics in the 21st Century
   Adrian E. Raftery, Martin A. Tanner, and Martin T. Wells (2001)
94 Accelerated Life Models: Modeling and Statistical Analysis
   Vilijandas Bagdonavicius and Mikhail Nikulin (2001)
95 Subset Selection in Regression, Second Edition *Alan Miller* (2002)
96 Topics in Modelling of Clustered Data
Marc Aerts, Helena Geys, Geert Molenberghs, and Louise M. Ryan (2002)
97 Components of Variance *D.R. Cox and P.J. Solomon* (2002)
98 Design and Analysis of Cross-Over Trials, 2nd Edition
   Byron Jones and Michael G. Kenward (2003)
99 Extreme Values in Finance, Telecommunications, and the Environment
   Bärbel Finkenstädt and Holger Rootzén (2003)
100 Statistical Inference and Simulation for Spatial Point Processes
   Jesper Møller and Rasmus Plenge Waagepetersen (2004)
101 Hierarchical Modeling and Analysis for Spatial Data
   Sudipto Banerjee, Bradley P. Carlin, and Alan E. Gelfand (2004)
102 Diagnostic Checks in Time Series *Wai Keung Li* (2004)
103 Stereology for Statisticians *Adrian Baddeley and Eva B. Vedel Jensen* (2004)
104 Gaussian Markov Random Fields: Theory and Applications
   Håvard Rue and Leonhard Held (2005)
105 Measurement Error in Nonlinear Models: A Modern Perspective, Second Edition
   *Raymond J. Carroll, David Ruppert, Leonard A. Stefanski,
   and Ciprian M. Crainiceanu* (2006)
106 Generalized Linear Models with Random Effects: Unified Analysis via H-likelihood
   Youngjo Lee, John A. Nelder, and Yudi Pawitan (2006)
107 Statistical Methods for Spatio-Temporal Systems
   Bärbel Finkenstädt, Leonhard Held, and Valerie Isham (2007)
108 Nonlinear Time Series: Semiparametric and Nonparametric Methods
   Jiti Gao (2007)
109 Missing Data in Longitudinal Studies: Strategies for Bayesian Modeling and Sensitivity Analysis
   Michael J. Daniels and Joseph W. Hogan (2008)
110 Hidden Markov Models for Time Series: An Introduction Using R
   Walter Zucchini and Iain L. MacDonald (2009)
111 ROC Curves for Continuous Data
   Wojtek J. Krzanowski and David J. Hand (2009)

Monographs on Statistics and Applied Probability 111

ROC Curves for Continuous Data

Wojtek J. Krzanowski
Imperial College
London, U. K.
and
University of Exeter
Exeter, U. K.

David J. Hand
Imperial College
London, U. K.

CRC Press
Taylor & Francis Group
Boca Raton   London   New York

CRC Press is an imprint of the
Taylor & Francis Group   an **informa** business

A CHAPMAN & HALL BOOK

Chapman & Hall/CRC
Taylor & Francis Group
6000 Broken Sound Parkway NW, Suite 300
Boca Raton, FL 33487-2742

© 2009 by Taylor and Francis Group, LLC
Chapman & Hall/CRC is an imprint of Taylor & Francis Group, an Informa business

No claim to original U.S. Government works

Printed in the United States of America on acid-free paper
10 9 8 7 6 5 4 3 2 1

International Standard Book Number: 978-1-4398-0021-8 (Hardback)

This book contains information obtained from authentic and highly regarded sources. Reasonable efforts have been made to publish reliable data and information, but the author and publisher cannot assume responsibility for the validity of all materials or the consequences of their use. The authors and publishers have attempted to trace the copyright holders of all material reproduced in this publication and apologize to copyright holders if permission to publish in this form has not been obtained. If any copyright material has not been acknowledged please write and let us know so we may rectify in any future reprint.

Except as permitted under U.S. Copyright Law, no part of this book may be reprinted, reproduced, transmitted, or utilized in any form by any electronic, mechanical, or other means, now known or hereafter invented, including photocopying, microfilming, and recording, or in any information storage or retrieval system, without written permission from the publishers.

For permission to photocopy or use material electronically from this work, please access www.copyright.com (http://www.copyright.com/) or contact the Copyright Clearance Center, Inc. (CCC), 222 Rosewood Drive, Danvers, MA 01923, 978-750-8400. CCC is a not-for-profit organization that provides licenses and registration for a variety of users. For organizations that have been granted a photocopy license by the CCC, a separate system of payment has been arranged.

Trademark Notice: Product or corporate names may be trademarks or registered trademarks, and are used only for identification and explanation without intent to infringe.

Library of Congress Cataloging-in-Publication Data

Krzanowski, W. J.
  ROC curves for continuous data / authors, Wojtek J. Krzanowski and David J. Hand.
    p. cm. -- (Monographs on statistics and applied probability ; 111)
  Includes bibliographical references and index.
  ISBN 978-1-4398-0021-8 (hard back : alk. paper)
  1. Receiver operating characteristic curves. I. Hand, D. J. II. Title. III. Series.

QA279.2.K73 2009
519.5'6--dc22   2009012820

Visit the Taylor & Francis Web site at
http://www.taylorandfrancis.com

and the CRC Press Web site at
http://www.crcpress.com

Contents

About the Authors	xi
Preface	xiii

Chapter 1: Introduction — 1
- 1.1 Background — 1
- 1.2 Classification — 2
- 1.3 Classifier performance assessment — 6
- 1.4 The ROC curve — 11
- 1.5 Further reading — 14

Chapter 2: Population ROC curves — 17
- 2.1 Introduction — 17
- 2.2 The ROC curve — 18
- 2.3 Slope of the ROC curve and optimality results — 24
- 2.4 Summary indices of the ROC curve — 25
- 2.5 The binormal model — 31
- 2.6 Further reading — 35

Chapter 3: Estimation — 37
- 3.1 Introduction — 37
- 3.2 Preliminaries: classification rule and error rates — 39
- 3.3 Estimation of ROC curves — 41
- 3.4 Sampling properties and confidence intervals — 57
- 3.5 Estimating summary indices — 63
- 3.6 Further reading — 74

Chapter 4: Further inference on single curves — 75
- 4.1 Introduction — 75
- 4.2 Tests of separation of P and N population scores — 76
- 4.3 Sample size calculations — 78
- 4.4 Errors in measurements — 82
- 4.5 Further reading — 86

Chapter 5: ROC curves and covariates — 87
- 5.1 Introduction — 87
- 5.2 Covariate adjustment of the ROC curve — 89
- 5.3 Covariate adjustment of summary statistics — 97

5.4 Incremental value	102
5.5 Matching in case-control studies	104
5.6 Further reading	105
Chapter 6: Comparing ROC curves	**107**
6.1 Introduction	107
6.2 Comparing summary statistics of two ROC curves	109
6.3 Comparing AUCs for two ROC curves	113
6.4 Comparing entire curves	115
6.5 Identifying where ROC curves differ	121
6.6 Further reading	122
Chapter 7: Bayesian methods	**123**
7.1 Introduction	123
7.2 General ROC analysis	125
7.3 Meta-analysis	128
7.4 Uncertain or unknown group labels	132
7.5 Further reading	139
Chapter 8: Beyond the basics	**141**
8.1 Introduction	141
8.2 Alternatives to ROC curves	141
8.3 Convex hull ROC curves	145
8.4 ROC curves for more than two classes	147
8.5 Other issues	154
8.6 Further reading	155
Chapter 9: Design and interpretation issues	**157**
9.1 Introduction	157
9.2 Missing values	158
9.3 Bias in ROC studies	165
9.4 Choice of optimum threshold	172
9.5 Medical imaging	176
9.6 Further reading	179
Chapter 10: Substantive applications	**181**
10.1 Introduction	181
10.2 Machine learning	182
10.3 Atmospheric sciences	184

10.4 Geosciences	190
10.5 Biosciences	192
10.6 Finance	195
10.7 Experimental psychology	198
10.8 Sociology	199
Appendix: ROC software	201
References	203
Index	227

About the Authors

Wojtek J. Krzanowski received a BSc in mathematics from Leeds University, a postgraduate diploma in mathematical statistics from Cambridge University, and a PhD in applied statistics from Reading University. He has been a scientific officer at Rothamsted Experimental Station in Harpenden; a senior research fellow at the RAF Institute of Aviation Medicine in Farnborough; successively lecturer, senior lecturer, and reader in applied statistics at the University of Reading; and professor of statistics at the University of Exeter. He retired from this post in 2005, and now holds the title of emeritus professor at Exeter as well as that of senior research investigator at Imperial College of Science, Technology and Medicine in London. His interests are in multivariate analysis, statistical modeling, classification, and computational methods. He has published 6 books, has made over 30 contributions to books, and has 100 articles in scientific journals to his credit. He is a former joint editor of the *Journal of the Royal Statistical Society, Series C*, a former associate editor of the *Journal of the Royal Statistical Society, Series B*, and has served on the editorial board of the *Journal of Classification* since its inception in 1984.

David J. Hand is head of the statistics section in the mathematics department, and head of the mathematics in banking and finance program of the Institute for Mathematical Sciences, both at Imperial College London. He studied mathematics at the University of Oxford and statistics and pattern recognition at the University of Southampton. His most recent books are *Information Generation: How Data Rule Our World* and *Statistics: A Very Short Introduction*. He launched the journal *Statistics and Computing*, and served a term as editor of the *Journal of the Royal Statistical Society, Series C*. He has been president of the International Federation of Classification Societies, and is currently president of the Royal Statistical Society. Among various awards for his research are the Guy Medal of the Royal Statistical Society, a Research Merit Award from the Royal Society, and the IEEE-ICDM Outstanding Contributions Award. He was elected a fellow of the British Academy in 2003. His research interests include classification methods, the fundamentals of statistics, and data mining. His applications interests include medicine, psychology, finance, and customer value management. He acts as a consultant to a wide range of organizations, including governments, banks, pharmaceutical companies, manufacturing industry, and health service providers.

Preface

The receiver operating characteristic (or ROC) curve is one of those statistical techniques that is now ubiquitous in a wide variety of substantive fields. Its earliest manifestation was during World War II for the analysis of radar signals, and it consequently entered the scientific literature in the 1950s in connection with signal detection theory and psychophysics (where assessment of human and animal detection of weak signals was of considerable interest). The seminal text for this early work was that by Green and Swets (1966). Later, in the 1970s and 1980s, it became evident that the technique was of considerable relevance to medical test evaluation and decision making, and the decades since then have seen much development and use of the technique in areas such as radiology, cardiology, clinical chemistry, and epidemiology. It was then but a short step before it permeated through a much wider range of applications, and it now features in subjects as diverse as sociology, experimental psychology, atmospheric and earth sciences, finance, machine learning, and data mining, among others.

While such widespread take-up of a statistical technique is gratifying for the statistician, a drawback is the equally wide dispersal of articles describing the latest advances. ROC curves feature heavily in textbooks, of course, but usually only as part of the story in a specific substantive area—such as the medical test evaluation and diagnosis texts by Zhou *et al.* (2002) and Pepe (2003)—and no single dedicated monograph or text seems to be available. So it is not easy for a biologist, say, to hunt down a relevant development that first arose in economics, or for a social researcher to come to grips with the technical details in a computer science journal.

Our purpose in writing this book, therefore, is to bring together in a single place all the relevant material for anyone interested in analyzing ROC curves, from whatever background they might come. We have included as much of the mathematical theory as we felt to be necessary without making the treatment indigestible, and we have included illustrative examples alongside all the major methodological developments. In the final chapter we also survey the uses made of the methodology across a range of substantive areas, and in the appendix we list a number of Web sites from which software implementing the various techniques described in earlier chapters can be downloaded. We therefore hope that the text will prove useful to a broad readership.

We would like to acknowledge the help of Ed Tricker and Dimitris Tasoulis in preparing the figures.

WJK
DJH

Chapter 1

Introduction

1.1 Background

A huge number of situations can be described by the following abstract framework. Each of a set of objects is known to belong to one of two classes. An assignment procedure assigns each object to a class on the basis of information observed about that object. Unfortunately, the assignment procedure is not perfect: errors are made, meaning that sometimes an object is assigned to an incorrect class. Because of this imperfection, we need to evaluate the quality of performance of the procedure. This might be so that we can decide if it is good enough for some purpose, attempt to improve it, replace it with another procedure, or for some other reason.

Examples of real problems which fit this abstract description include:

1. conducting medical diagnosis, in which the aim is to assign each patient to disease A or disease B;

2. developing speech recognition systems, in which the aim is to classify spoken words;

3. evaluating financial credit applications, in which the aim is to assign each applicant to a "likely to default" or "not likely to default" class;

4. assessing applicants for a university course, on the basis of whether or not they are likely to be able to pass the final examination;

5. filtering incoming email messages, to decide if they are spam or genuine messages;

6. examining credit card transactions, to decide if they are fraudulent or not;

7. investigating patterns of gene expression in microarray data, to see if they correspond to cancer or not.

In fact, the list of situations which fall into the above abstract description is essentially unlimited. In some cases, of course, more than two classes might be involved, but the case of two classes is by far the most important one in practice (sick/well, yes/no, right/wrong, accept/reject, act/do not act, condition present/absent, and so on). Moreover, the multi-class case can often be decomposed into a sequence of two-class cases, and we say more about this below.

There are various ways in which the quality of performance of such systems may be evaluated. A brief overview is given in Section 3. This book is concerned with one particular, but extremely important and widely used approach. This is the *Receiver Operating Characteristic* (ROC) curve, defined below. The name Receiver Operating Characteristic arises from the use of such curves in signal detection theory (Green and Swets (1966), Egan (1975)), where the aim is to detect the presence of a particular signal, missing as few genuine occurrences as possible while simultaneously raising as few false alarms as possible. That is, in signal detection theory the aim is to assign each event either into the signal class or into the nonsignal class—so that the abstract situation is the same as above. The word "characteristic" in Receiver Operating Characteristic refers to the characteristics of behavior of the classifier over the potential range of its operation.

1.2 Classification

The information about each object which is used to assign it to a class can be regarded as a vector of descriptive variables, characteristics, or features. The type of information that one obtains depends on the level of measurement of each variable: a *nominal* variable is one whose "values" are categories (e.g., color of eyes); a *binary* variable is a nominal variable having just two possible categories (e.g., presence or absence of pain); an *ordinal* variable has categories that are ordered in some way (e.g., no pain, mild pain, moderate pain, severe pain); a *discrete* (numerical) variable is one that can take only a finite number of distinct possible values (e.g., the number of students who wear glasses in a class of 25); and a *continuous* (numerical) variable is one that can take any value in either a finite or infinite range (e.g., the weight of a patient in hospital). Sometimes the measuring device limits the possible values that a continuous variable can take (e.g., if the weighing machine only gives weights to the nearest gram), but we generally treat the resultant variable as continuous rather than discrete as this is its underlying characteristic.

Sometimes the vector of descriptive variables will be univariate—it will consist of just a single variable—but often it will be multivariate. In the single variable case, that variable may be regarded as a proxy for or approximation to some other variable which defines the classes. Thus, just to take an example, in a medical context, the single variable might be erythrocyte sedimentation rate. This is a measure of inflammation often used in medical screening. Inflammation causes red blood cells to stick together, so that they tend to fall faster. Our aim will be to assign each patient to ill/not ill classes on the basis of their response to this test.

In the more general multivariate context, each object will have a vector $\boldsymbol{X}$ of measurements. Again in a medical context, we might seek to assign each of a population of subjects to the classes "will/will not develop osteoporosis in later life" on the basis of responses to a questionnaire describing their lifestyles, diet, and medical history: each item in the questionnaire corresponds to one measurement. In general, multivariate methods are more powerful than univariate methods, if only because each component of the descriptive vector can add extra information about the class of the object. The multiple measurements taken on each object are then reduced to a single score $S(\boldsymbol{X})$ for that object by some appropriate function. We say more about the possible functions S below. Like the individual variables, the scores for each object can be of any of the previously defined measurement types. However, the majority of functions S with which we are typically concerned will convert the raw information into a continuous value. Hence for the purposes of this book we will assume that ultimately, whether single or multiple pieces of information are captured for each object, each object is assigned a score on a univariate continuum. The class assignment or classification is then made by comparing this score with a threshold: if the score is above the threshold they are assigned to one class, and if the score is below the threshold to the other. Objects with scores which are precisely equal to the threshold, not a common occurrence for continuous data but one which can arise with certain kinds of function S, can be assigned arbitrarily. We denote the characteristics describing objects as $\boldsymbol{X}$, with $\boldsymbol{x}$ denoting particular values, and the resulting scores as $S(\boldsymbol{X})$, taking particular values $s(\boldsymbol{x})$. The classification threshold T takes values denoted by t.

We will denote the two classes by P and N. In some cases there is a symmetry between the two populations. An example might be the attempt to assign beetles to one of two species on the basis of measurements on their carapaces, and neither species plays a dominant or preferred role. In other cases, however, there will be an asymmetry between the two populations. For example, in trying to assign patients to "will/will not recover if treated" classes, there is clearly asymmetry. Our P, N notation allows us to represent this asymmetry—with P being read as "positive" (cases, people with the disease, fraudulent transactions, etc.) and N as "negative" (noncases, normals, healthy people, legitimate transactions, etc.) Generally, the emphasis is on identifying P individuals correctly.

Because the abstract classification problem outlined above is so ubiquitous, a great amount of research has been carried out on it. As a consequence, a very large number of methods for developing classification rules have been developed. Moreover, again because of the ubiquity of the problem, such methods have been developed by a wide range of research communities, including people working in statistics, pattern recognition, data mining, and machine learning, and also including many working in specialized application areas such as medicine and speech recognition. It probably does not need remarking that particular highly refined methods have been developed for particular domains. The area of biometrics is an example, with methods being developed for classifying faces, fingerprints, DNA traces, iris and retinal patterns, gait, and other aspects of individuals.

A central aspect to developing classification rules is to choose the function S which reduces the vector $\boldsymbol{x}$ to a single score. There is a very extensive literature describing such functions, with formal development dating back to the early part of the twentieth century and with the advent of the computer having led to both major practical advances and major theoretical advances. Some general introductions to the topic are listed at the end of this chapter, and here we content ourselves with merely briefly describing some basic ideas.

The fundamental aim is to construct a score function $S(\boldsymbol{X})$ such that members of the two classes have distinctly different sets of scores, thereby enabling the classes to be clearly distinguished. We will assume that the scores have been orientated in such a way that members of class P tend to have large scores and members of class N tend to have small scores, with a threshold that divides the scores into two groups. A common situation is that one has a "training set" or "design set" of data describing previous objects, including both the descriptive vectors $\boldsymbol{X}$ and the true P or N classes of each of the objects in this set. Any proposed function S will then produce a distribution of scores for the members of P in the training set, and a distribution of scores for the members of N in the training set, and the score for any particular object can then be compared with the classification threshold t. A good function S and choice of classification threshold will tend to produce scores above the threshold for members of P and below it for members of N.

There are two distinct aspects to constructing such rules. The first is the form of the rule: is it a weighted sum of the raw components of $\boldsymbol{X}$, a partition of the multivariate $\boldsymbol{X}$ space, a sum of nonlinear transformations of the components of $\boldsymbol{X}$, or any of a host of other ways of combining these components? The second is the criterion used to estimate any parameters in S. In a weighted sum, for example, we must choose the weights, in a partition of $\boldsymbol{X}$ we must choose the positions of the cutpoints, and so on. Some criterion must also be chosen to compare and choose between different functions. This question of choice of optimization criterion is the central theme of this book, and is described in the next section.

In the situation described above we supposed we had a data set which not only had the descriptive vectors for a set of objects, but also had their true class labels. Such a situation is sometimes called "supervised classification" because it is as if a "supervisor" was providing the true class labels. However, this is not the only sort of data from which one may need to construct a classification rule. Another situation which sometimes arises is when one is seeking to assign objects to the classes "standard" or "nonstandard." One may have a sample of objects from the former class, but no objects from the latter. Using these data one can construct a measure of how typical or "representative" any object is of the standard class. Again a threshold is adopted, so that any new object yielding a score which is very unusual is classified as "nonstandard."

Note that in all of this the true classes may exist but simply not be observable at the time of applying the rule, or they may not actually exist. An example of the former is a medical condition whose presence cannot be definitively established without a dangerous invasive procedure. The alternative may be a classification rule based on simple and quick tests. Eventually the true class does become known (perhaps post mortem), and a set of such patients can form the training set. An example of the latter, where the true class does not actually exist, arises in prognostic medicine, in which we are seeking to identify those patients who *will* develop a particular disease if they are not treated. Once again, a sample of untreated patients will provide cases who eventually do and do not develop the disease and these can form the training set.

The training data are used to construct the classification function S. The rule will then be used to classify new objects. With this in mind, one clearly wishes the training data to be "representative" of the population to which the rule will be applied. One would have little hope of successfully diagnosing patients as suffering from disease A versus disease B if the training sample had cases with completely different diseases. This, of course, means that care must be taken in choosing the training set, and also in the statements one makes about the applicability of any classification rule. Samples drawn from the general population are likely to be different from samples drawn from specialized clinics.

1.3 Classifier performance assessment

Training data are used to construct the classification rule. Then, having finally settled on a rule, we want to know how effective it will be in assigning future objects to classes. To explore this, we need actually to assign some objects to classes and see, in some way, how well the rule does. In principle, one could simply apply the rule to the training set, and examine the accuracy with which it classifies those objects. This, however, would be unwise. Since the classification rule has been constructed using the training set it is, in some sense, optimized for those data. After all, since the training data are a sample from the popula-

1.3. CLASSIFIER PERFORMANCE ASSESSMENT

tion one is seeking to classify, it would be perverse to have deliberately chosen a classification rule which performed badly on those data. But if the rule has been chosen because it does well on the training data there is a real possibility that the performance on those data will be better than on another sample drawn from the same distribution. We thus require more subtle assessment methods.

Various approaches have been developed, all hinging around the notion of separating the data set used for constructing the rule from the data set used to evaluate it. The simplest approach is to divide the available data into two sets, a training and test set, the former for choosing the rule and the latter for assessing its performance. This division can be repeated, splitting the data in multiple ways, and averaging the results to avoid misleading estimates arising from the chance way the data are split. Other methods split the data in unbalanced ways, at an extreme putting only one data point in the test set and the others all in the training set, and then repeating this for each available data point, again averaging the results. This is the *leave-one-out* method. Yet other approaches draw random sets with replacement from the available data, and use each of these as training sets, allowing estimates of the optimism of reclassifying the overall training set to be obtained. These are *bootstrap* methods. Because of its importance, considerable attention has been devoted to this problem, but this book is not the appropriate place to discuss such methods. References to extended discussions are given at the end of this chapter.

Finding a way to obtain unbiased estimates of future performance is important, but we have not yet discussed how performance is actually measured. To discuss this, we need to establish a framework and some terminology.

We have already seen that, for each object, a classification rule will yield a score $s(\boldsymbol{X})$ on a univariate continuum. For objects in the "positive" group, P, this will result in a distribution of scores $p(s|\text{P})$ and, for objects in the "negative" group, N, a distribution $p(s|\text{N})$. Classifications are given by comparing the scores with a threshold T. If we can find a threshold $T = t$ such that all members of class P have scores that are all greater than t, and all members of class N have scores that are all less than or equal to t, then perfect classification can be attained. But this is an unusual situation. Typically, the two sets of scores overlap to some extent and perfect classification is impossible. When this is the case, performance is measured by the extent to which scores for objects in class P tend to take large values, and scores for objects in class N tend to take small values. And it will be immediately clear from the use of expressions such as "the extent to which" and "tend to take" in the above that there will be various ways in which these ideas can be made concrete. Many such methods (and in particular the class of methods which is the concern of this book) are based on the two-by-two classification table which results from cross-classifying the true class of each object by its predicted class. For a test set, the proportions of the test set which fall in the cells of this table are empirical realizations of the joint probabilities $p(s > t, P)$, $p(s > t, N)$, $p(s \leq t, P)$, $p(s \leq t, N)$. Different ways of summarizing these four joint probabilities yield different measures of classification performance.

One very common measure is the misclassification or error rate: the probability of a class N object having a score greater than t or a class P object having a score less than t. In fact, in a meta-analysis of comparative studies of classification rules, Jamain (2004) found that over 95% of such studies based their conclusions on misclassification rate.

But misclassification rate is far from perfect. At one level, this must be obvious. The four probabilities listed above provide three independent pieces of information, so no single summary can capture everything in them. At a more mundane level, however, misclassification rate might be inappropriate because it weights the two kinds of misclassification (class N misclassified as P, and vice versa) as equally important. Misclassifying as healthy someone suffering from a potentially fatal disease, which can be easily cured with an otherwise harmless treatment if detected in time, is likely to be more serious than the converse misclassification. Misclassification rate fails to take this difference into account. This weakness can be overcome by weighting the misclassifications according to their perceived severity. However, instead of going into further details, this book will develop things in a rather more general direction, of which weighted misclassification rate can be seen to be a special case.

We can conveniently summarize the four joint probabilities above in terms of two conditional probabilities and one marginal probability:

1. the probability that an object from class N yields a score greater than t: $p(s > t|N)$; this is the *false positive rate*, denoted *fp*;

2. the probability that an object from class P yields a score greater

1.3. CLASSIFIER PERFORMANCE ASSESSMENT

than t: $p(s > t|P)$; this is the *true positive rate*, denoted tp;

3. the marginal probability that an object belongs to class P: $p(P)$.

A false positive thus arises when an object which really belongs to class N is incorrectly assigned to class P, because its score falls above the threshold t. A true positive arises when an object which really belongs to class P is correctly assigned to class P, again because its score falls above t.

There are also two complementary conditional rates, and one complementary marginal probability:

1. the true negative rate, $p(s \leq t|N)$, the proportion of class N objects which are correctly classified as class N, equal to $1 - fp$ and denoted tn;

2. the false negative rate, $p(s \leq t|P)$, the proportion of class P objects which are incorrectly classified as class N, equal to $1 - tp$ and denoted fn.

3. the marginal probability that an object belongs to class N: $p(N) = 1 - p(P)$.

Epidemiologists sometimes use the term *Sensitivity*, denoted Se, to mean the same as our true positive rate, and *Specificity*, denoted Sp, to mean the same as our true negative rate. They also use the term *prevalence* to mean the proportion of a population which has a disease; $p(P)$ if "positive" corresponds to the disease state.

The true/false positive/negative rates and the specificity and sensitivity are all conditional probabilities of having a particular predicted class given the true class. We can also define the converse conditional probabilities of having a particular true class given the predicted class. Thus the *positive predictive value*, ppv, of a classification method is the proportion of objects which really are class P amongst all the objects which the rule assigns to class P: $p(P|s > t)$. The *negative predictive value*, npv, is the proportion of objects which are really class N amongst all the objects which the classification rule assigns to class N: $p(N|s \leq t)$.

Other disciplines, having developed the ideas independently, use different names for the same or similar concepts. Thus, for example, in the information retrieval and data mining literature, *recall* is the ratio of the number of relevant records retrieved to the total number of relevant records in the database, and *precision* is the ratio of the number of relevant records retrieved to the total number of records retrieved. If we let "relevant" correspond to class P, then these ratios are the same as the sample *tp* and *ppv* values respectively.

Incidentally, we should point out in the interest of avoiding confusion that the definitions of false positive and false negative rate given above, while used by almost all authors, are not quite universally used. On page 4 of Fleiss (1981), for example, he defines the false positive rate as the proportion of those objects which have a score above the threshold (i.e., are predicted as being positive) which are really negative; that is, as the complement of our *ppv*. The reader needs to check which definitions are used when reading other accounts and, perhaps, adopt the standard definitions used in this account when writing material.

There are obvious and straightforward relationships between the various conditional, marginal, and joint probabilities, and these can sometimes be useful in switching from one perspective on the performance of a test to another. For example, the misclassification rate e of a classification rule can be expressed as a weighted sum of the true positive and false positive rate: $e = (1 - tp) \times p(\text{P}) + fp \times p(\text{N})$.

Note that the true positive rate and the true negative rate have the property of being independent of the marginal probabilities $p(\text{N}), p(\text{P})$, which are also known as the *class priors*. This gives them power to generalize over populations which have different proportions belonging to the classes, and is a property we will make use of below when describing the ROC curve. Having said that, since, once again, three joint probabilities are needed to capture the complete properties of the classification table, the pair of true positive and true negative rates alone will not be sufficient. This is nicely illustrated by the case of rare diseases. Suppose we have developed a classification rule which correctly classifies 99% of the positives and also correctly classifies 99% of the negatives. That is, the rule's true positive and true negative values are both 0.99. At first glance, this looks like an excellent rule—in few applications are such large values achieved. But now suppose that the true proportion of positives in the population is only 1 in a 1000: we are dealing with a rare disease. Then, using Bayes' theorem, we have

$$p(\text{P}|s > t) = \frac{p(s > t|\text{P}) p(\text{P})}{p(s > t|\text{P}) p(\text{P}) + p(s > t|\text{N}) p(\text{N})}$$

from which $p(\text{P}|s > t) = 0.09$. That is, amongst all those which the test predicts to be positive, only 9% are really positive. A real example of this phenomenon is given in Williams *et al.* (1982) for anorexia nervosa.

The true positive and true negative rates of a classification rule are usually used together as joint measures of performance. This is because they are complementary: in general, decreasing t so that the true positive rate increases will lead to the true negative rate decreasing. Some acceptable compromise has to be reached. Misclassification rate is one proposal for such a compromise: it is that value of t which yields the overall minimum of the weighted sum $e = (1 - tp) \times p\,(\mathrm{P}) + fp \times p\,(\mathrm{N})$. If the misclassifications have different degrees of severity, as described above, then different choices of weights (no longer simply $p(\mathrm{P})$ and $p(\mathrm{N})$) can be used to minimize the overall loss. Yet another common proposal is to choose the threshold to maximize $tp - fp$, equivalently $tp + tn - 1$ or Sensitivity + Specificity -1. The maximum value of this quantity is the *Youden index*, YI.

It should be clear that the essence of all these proposed performance measures, and indeed others not yet mentioned, is that they are based on some sort of comparison between the distributions of the scores for the positive and negative populations. As we have already commented, a good rule tends to produce high scores for the positive population, and low scores for the negative population, and the classifier is better the larger the extent to which these distributions differ. The ROC curve, to be introduced in the next section, is a way of jointly displaying these two distributions. By suitably interpreting the curve, any measure of performance based on these distributions can be seen. As a final illustration of this, we refer to just one more measure: the *Area Under the Curve* or *AUC*. As we will see in later chapters, the AUC, based on the ROC curve, is a global measure of separability between the distributions of scores for the positive and negative populations. It does not require one to choose a threshold value, but summarizes the results over all possible choices.

1.4 The ROC curve

As the previous section demonstrated, there are many possible ways to measure the performance of classification rules. Ideally, one should choose that particular performance criterion which corresponds to the aspect of performance that one considers to be central to the particular application. Often, however, it is difficult to identify one single such central aspect, and in any case one may not know the precise circumstances under which the rule will be applied in the future—circumstances do change, after all. For such reasons it is often helpful to have a way of displaying and summarizing performance over a wide range of conditions. And this is exactly what the ROC curve does.

The ROC curve is a graph showing true positive rate on the vertical axis and false positive rate on the horizontal axis, as the classification threshold t varies. It is a single curve summarizing the information in the cumulative distribution functions of the scores of the two classes. One can think of it as a complete representation of classifier performance, as the choice of the classification threshold t varies. The implications of this plot, its interpretation, and how a great many performance measures can be read from it are described in the next chapter. For example, misclassification rate will be seen to be simply the minimum distance between the curve and the upper left corner of the square containing the ROC plot, when the axes are appropriately weighted by the class sizes. And the area under the curve will turn out to be equivalent to the popular Mann-Whitney U-statistic: it is a measure of the similarity of the two score distributions. The relationship between the ROC curve and these and many other measures are described in Chapter 2.

Chapter 3 then extends the theory into practice by describing how ROC curves are estimated. Many approaches have been developed, including nonparametric methods based directly on the empirical distribution functions of scores for the two classes, and parametric methods based on assuming distributional forms for the scores. The *binormal* form is a particularly important special case. Since summary statistics based on ROC curves provide key insights into the performance of classification rules it is often useful to estimate these statistics directly, and this chapter also discusses such estimation.

Chapter 4 builds on the theory of the earlier chapters and extends the discussion into the realm of inference, covering statistical tests of ROC curves and their summary statistics.

Up to this point, the book has supposed that each classification rule produces a single ROC curve. Sometimes, however, the rule will be parameterized by the values of one or more covariates, with different values producing different curves, so yielding a whole family of curves. Chapter 5 extends the discussion to consider the impact of covariates on ROC curves. It surveys the various methods that have been derived for adjusting the curve and its summary statistics to take account of different covariate values, and illustrates the use of these methods in practice.

Chapter 6 then looks at the very important special problem of comparing two ROC curves. This is generally a crucial aspect, since often one wants to know not merely how a given classification tool performs, but also how different tools compare. Is one better than another or, more generally, under what circumstances is one better than another? Often one finds that one method is superior for some values of the classification threshold but inferior for others. This manifests itself by ROC curves which cross over. Moreover, very often discussions of multiple curves are based on the same underlying objects, so that correlated curves result. Inference in such cases requires care.

Most of the methodology associated with ROC curves was, until fairly recently, developed within the classical (or "frequentist") framework of statistical inference, so this is the framework behind the methods considered in Chapters 2 to 6. However, following the major computational advances that were made in the 1990s, the Bayesian approach to inference has become very popular and many Bayesian techniques now exist alongside the classical techniques in most branches of statistics. ROC analysis is no exception, although perhaps the development of Bayesian methods has been slower here than in some other areas. Nevertheless, various Bayesian methods do now exist and they are described in Chapter 7.

Chapters 2 to 7 thus contain what might be termed the fundamental theory of ROC curves, and the remaining chapters are concerned with various special topics. Chapter 8 considers extensions of the basic analysis to cope with more complex situations, such as the combination of multiple ROC curves and problems induced by the presence of more than two classes. We have already briefly mentioned this last situation, which potentially presents some formidable dimensionality problems if a single overall analysis is sought. However, a simplified approach is to treat the situation by a series of two-class analyses, and this can be achieved in various ways. One possibility when there are k classes is to produce k different ROC curves by considering each class in turn as population P and the union of all other classes as population N, while another is to produce all $k(k-1)$ distinct pairwise-class ROC curves. Both approaches can be used to derive suitable generalizations of summary statistics such as the AUC.

Chapter 9 is then concerned with design and interpretation issues, covering topics such as missing data, verification bias, sample size determination, design of ROC studies, and choice of optimum threshold from the ROC curve. Finally, Chapter 10 describes a range of substantive applications, not only to illustrate some of the techniques whose theory has been developed in earlier chapters but also to demonstrate the very wide applicability of these techniques across different subject areas.

In view of this last sentence, it seems appropriate to conclude the chapter with some data illustrating the growth of interest that has taken place in ROC curves and their analysis since they first appeared. Earliest references to the topic can be traced back to the 1950s, and for the next 30 years there was a slow but steady annual increase in the number of articles that had some aspect of ROC analysis as a central feature. This growth can be attributed partly to usage of ROC curves in the historical areas of signal detection and psychophysics, and partly to methodological developments of the associated techniques. However, some time in the 1980s ROC techniques became established analytical tools in a variety of disciplines focussed on diagnosis of one form or another (such as radiography or credit scoring, for example), and the rate of increase of publications accelerated dramatically. To illustrate the trend, we conducted a search of journals in science, the social sciences, and the arts and humanities, via the ISI Web of Knowledge, using the key phrases "receiver operating characteristic" and "ROC curves." The numbers of article that were flagged as containing one or other of these phrases is shown in Table 1.1, for all 4-year periods from the late 1950s to 2007.

The sudden jump in interest around 1990, and the exponential rate of growth since then, is very evident. Furthermore, the number of articles flagged for the first ten months of 2008 stands at 1461, so it is clear that this explosive growth is set to continue in the immediate future. We therefore hope that the present book is a timely one that will be of service to both present and future users of ROC analysis.

1.5 Further reading

Some general references on developing classification rules are McLachlan (1992), Ripley (1996), and Webb (2002). Hand (1997 and 2001)

1.5. FURTHER READING

Table 1.1: Articles relating to ROC analysis.

Dates	No. of articles
Pre-1964	2
1964 – 67	7
1968 – 71	8
1972 – 75	9
1976 – 79	18
1980 – 83	29
1984 – 87	41
1988 – 91	192
1992 – 95	854
1996 – 99	1582
2000 – 03	2506
2004 – 07	4463

provide overviews of criteria for estimating the performance of classification methods, while Hand (1986) and Schiavo and Hand (2000) describe estimation of misclassification rate.

The ROC curve has a long history, and has been rediscovered many times by researchers in different disciplines. This is perhaps indicative of its extreme usefulness, as well as of its naturalness as a representation of a classification rule. Important earlier references include work in electronic signal detection theory, psychology, and medicine—see, for example, Peterson, Birdsall, and Fox (1954), Green and Swets (1966), Bamber (1975), Egan (1975), Metz (1978), Swets and Pickett (1982), Hanley and McNeil (1982), and Hanley (1989). More recently, extensive discussions in a medical context have been given by Zhou *et al.* (2002) and Pepe (2003).

Chapter 2

Population ROC curves

2.1 Introduction

In this chapter we focus on the theoretical concepts that underly ROC analysis, and provide the necessary framework within which statistical methods can be developed. Given the background ideas presented in Chapter 1, we take as our starting point the existence of two populations—a "positive" population P and a "negative" population N—together with a classification rule for allocating unlabelled individuals to one or other of these populations. We assume this classification rule to be some continuous function $S(\boldsymbol{X})$ of the random vector $\boldsymbol{X}$ of variables measured on each individual, conventionally arranged so that large values of the function are more indicative of population P and small ones more indicative of population N. Thus if $\boldsymbol{x}$ is the observed value of $\boldsymbol{X}$ for a particular individual and $s(\boldsymbol{x})$ is the function score for this individual, then the individual is allocated to population P or population N according as $s(\boldsymbol{x})$ exceeds or does not exceed some threshold T.

Theoretical development is based upon probability models of the two populations. Since the classification function $S(\boldsymbol{X})$ is the determining component of the analysis, the relevant probability models are those for the classification scores obtained from the random vector of initial measurements $\boldsymbol{X}$. Let us therefore write, quite generally, $p(s|\text{P})$ and $p(s|\text{N})$ for the probability density functions of the classification scores pertaining to individuals coming from populations P and N respectively. In order to calculate any probabilities for a population, specific distributional forms have to be assumed for these density functions. This can be done directly for the scores, on the basis perhaps of either prior beliefs or previous experience about their behavior. Alternatively, it can be done by assuming particular distributions for the original measurements X in each population and then deriving the distributions of the scores by mathematical manipulation. We will consider a specific model near the end of this chapter, but for the present we maintain generality; it is worth noting, however, that while the distributions of the original measurements are multivariate those for the scores are (usually) univariate and hence mathematically simpler.

Our aim in this chapter is to define the ROC curve for such models, to establish some of its chief properties, to highlight features of the curve that are of practical interest and how they are related to other familiar population quantities, and to derive the most useful special cases of the general formulation. The question of data analysis and statistical inference is left to succeeding chapters.

2.2 The ROC curve

2.2.1 Definition

Suppose that t is the value of the threshold T in a particular classification rule, so that an individual is allocated to population P if its classification score s exceeds t and otherwise to population N. In order to assess the efficacy of this classifier we need to calculate the probability of making an incorrect allocation. Such a probability tells us the *rate* at which future individuals requiring classification will be misallocated. More specifically, as already mentioned in Chapter 1, we can define four probabilities and their associated rates for the classifier:

1. the probability that an individual from P is correctly classified, i.e., the true positive rate $tp = p(s > t|\text{P})$;

2. the probability that an individual from N is misclassified, i.e., the false positive rate $fp = p(s > t|\text{N})$;

3. the probability that an individual from N is correctly classified, i.e., the true negative rate $tn = p(s \leq t|\text{N})$; and

4. the probability that an individual from P is misclassified, i.e., the false negative rate $fn = p(s \leq t|\text{P})$.

2.2. THE ROC CURVE

Given probability densities $p(s|\text{P})$, $p(s|\text{N})$, and the value t, numerical values lying between 0 and 1 can be obtained readily for these four rates and this gives a full description of the performance of the classifier. Clearly, for good performance, we require high "true" and low "false" rates.

However, this is for a particular choice of threshold t, and the best choice of t is not generally known in advance but must be determined as part of the classifier construction. Varying t and evaluating all the four quantities above will clearly give full information on which to base this decision and hence to assess the performance of the classifier, but since $tp + fn = 1$ and $fp + tn = 1$ we do not need so much information. The ROC curve provides a much more easily digestible summary. It is the curve obtained on varying t, but using just the true and false positive rates and plotting (fp, tp) as points against orthogonal axes. Here fp is the value on the horizontal axis (abscissa) and tp is the value on the vertical axis (ordinate). Figure 2.1 shows three such curves.

2.2.2 General features

The purpose of the ROC, as indicated above, is to provide an assessment of the classifier over the whole range of potential t values rather than at just a single chosen one. Clearly, the worth of a classifier can be judged by the extent to which the two distributions of its scores $p(s|\text{P})$ and $p(s|\text{N})$ differ. The more they differ, the less in general will there be any overlap between them so the less likely are incorrect allocations to be made, and hence the more successful is the classifier in making correct decisions. Conversely, the more alike are the two distributions, the more overlap there is between them and so the more likely are incorrect allocations to be made.

Let us consider the extremes. The classifier will be least successful when the two populations are exactly the same, so that $p(s|\text{P}) = p(s|\text{N}) = p(s)$, say. In such a case the probability of allocating an individual to population P is the same whether that individual has come from P or from N, the exact value of this probability depending on the threshold value t. So in this case, as t varies tp will always equal fp and the ROC curve will be the straight line joining points $(0,0)$ and $(1,1)$. This line is usually called the *chance diagonal*, as it represents essentially random allocation of individuals to one of the two populations.

At the other (usually unattainable) extreme there is complete separation of $p(s|\text{P})$ and $p(s|\text{N})$. In this case there will be at least one value

t at which perfect allocation of each individual is achieved, so for such t we have $tp = 1$ and $fp = 0$. Moreover, since the ROC curve focusses only on the probabilities that $s > t$ in the two populations, then for all smaller values of t we must have $tp = 1$ while fp varies from 0 to 1 and for all larger values of t we must have $fp = 0$ while tp varies from 1 to 0. So the ROC must therefore lie along the upper borders of the graph: a straight line from $(0,0)$ to $(0,1)$ followed by a straight line from $(0,1)$ to $(1,1)$.

In practice, the ROC curve will be a continuous curve lying between these two extremes, so it will lie in the upper triangle of the graph. The closer it comes to the top left-hand corner of the graph, the closer do we approach a situation of complete separation between populations, and hence the better is the performance of the classifier. Note that if a ROC curve lies in the lower triangle, then this simply indicates that the score distribution has the wrong orientation and a reversal is needed (with allocation to P indicated if $s < t$ rather than $s > t$). More specifically, if just some points lie in the lower triangle then a reversal for the t values corresponding to those points will correct matters. There is thus no loss in generality in assuming that the curve will always lie in the upper triangle of the graph.

Figure 2.1 shows three such curves plus the chance diagonal. The solid curve corresponds to the best classifier, because at any fp value it has higher tp than all the others while at any tp value it has lower fp than all the others.

2.2.3 Properties of the ROC

To study some of the main properties, let us define the ROC in more familiar mathematical notation as the curve $y = h(x)$, where y is the true positive rate tp and x is the false positive rate fp. Since the points (x, y) on the curve are determined by the threshold T of the classification score S, x and y can be written more precisely as functions of the parameter t, viz. $x(t) = p(s > t|N)$ and $y(t) = p(s > t|P)$. However, we will only use this expanded notation if the presence of this parameter needs to be emphasized.

Property 1

$y = h(x)$ is a monotone increasing function in the positive quadrant, lying between $y = 0$ at $x = 0$ and $y = 1$ at $x = 1$.

2.2. THE ROC CURVE

Figure 2.1: Three ROC curves, plus chance diagonal

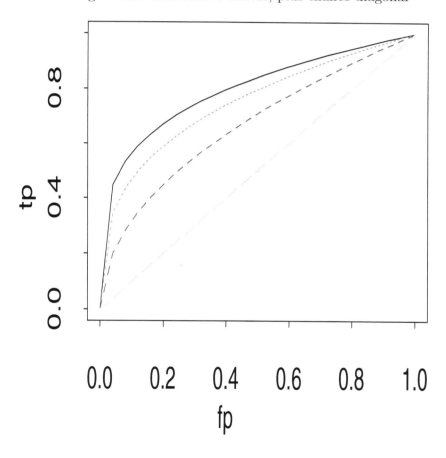

Proof: Consideration of the way that the classifier scores are arranged shows that both $x(t)$ and $y(t)$ increase and decrease together as t varies. Moreover, $\lim_{t\to\infty} x(t) = \lim_{t\to\infty} y(t) = 0$ and $\lim_{t\to-\infty} x(t) = \lim_{t\to-\infty} y(t) = 1$, which establishes the result.

Property 2

The ROC curve is unaltered if the classification scores undergo a strictly increasing transformation.

Proof: Suppose that $U = \phi(S)$ is a strictly increasing transformation, i.e., $S_1 > S_2 \Leftrightarrow U_1 = \phi(S_1) > U_2 = \phi(S_2)$. Consider the point on the ROC curve for S at threshold value t, and let $v = \phi(t)$. Then it follows that
$$p(U > v|\text{P}) = p(\phi(S) > \phi(t)|\text{P}) = p(S > t|\text{P})$$
and
$$p(U > v|\text{N}) = p(\phi(S) > \phi(t)|\text{N}) = p(S > t|\text{N})$$
so that the same point exists on the ROC curve for U. Applying the reverse argument to each point on the ROC curve for U establishes that the two curves are identical.

Property 3

Providing that the slope of the ROC at the point with threshold value t is well-defined, it is given by
$$\frac{dy}{dx} = \frac{p(t|\text{P})}{p(t|\text{N})}.$$

Proof: First note that
$$y(t) = p(S > t|\text{P}) = 1 - \int_{-\infty}^{t} p(s|\text{P})ds,$$
so that
$$\frac{dy}{dt} = -p(t|\text{P}).$$
Thus
$$\frac{dy}{dx} = \frac{dy}{dt}\frac{dt}{dx} = -p(t|\text{P})\frac{dt}{dx}.$$

2.2. THE ROC CURVE

Moreover,
$$x(t) = p(S > t|N) = 1 - \int_{-\infty}^{t} p(s|N)ds,$$

so that
$$\frac{dx}{dt} = -p(t|N).$$

Also
$$\frac{dt}{dx} = 1 \bigg/ \frac{dx}{dt},$$

which establishes the result.

Property 3 leads to some interesting connections between the slope of the ROC curve and other familiar statistics, as given in Section 2.3 below.

2.2.4 Continuous scores

All the development in this chapter is perfectly general, in that it can be applied to any type of measurement scale for the scores S. However, since our focus in the book is specifically on continuous scores we here provide a more convenient form for the ROC curve in this case, and will make use of this form in most of the later chapters.

From the previous section we see that the ROC curve is the curve $y(t) = h(x(t))$ plotted over values of t, where $x(t) = p(S > t|N)$ and $y(t) = p(S > t|P)$. When S is a continuous variable it will possess density and distribution functions in each of the two populations, so let us denote its density and distribution functions in population N by f, F respectively, and in population P by g, G respectively. Thus $p(s|N) = f(s), p(s|P) = g(s), x(t) = 1 - F(t)$ and $y(t) = 1 - G(t)$. Eliminating t from this last pair of equations using standard mathematics yields

$$y = 1 - G[F^{-1}(1-x)] \quad (0 \le x \le 1),$$

and this will be the most convenient equation of the ROC curve to use with continuous scores.

2.3 Slope of the ROC curve and optimality results

There are two main results of interest, one from decision theory and the other from the theory of hypothesis testing.

2.3.1 Cost-weighted misclassification rates

Classification errors are incurred when we allocate an N individual to P or a P individual to N, and if threshold t is used on the classifier scores then their respective probabilities are $x(t)$ and $1-y(t)$. Suppose that the costs (however defined) of making these errors are $c(P|N)$ and $c(N|P)$, and that the relative proportions of P and N individuals in the population are q and $1-q$ respectively. Then the expected cost due to misclassification when using the classifier threshold t is

$$C = q[1 - y(t)]c(N|P) + (1-q)x(t)c(P|N).$$

Provided that the slope of the ROC curve is monotonically decreasing, the threshold that minimizes this cost is given by solving the equation

$$\frac{dC}{dx} = 0,$$

and straightforward differentiation shows this to be the threshold corresponding to slope

$$\frac{dy}{dx} = \frac{(1-q)c(P|N)}{qc(N|P)}$$

of the ROC curve. We will return to the question of differential costs of misclassification from time to time. In the case when the two costs are equal and also when the two population rates are equal (i.e., $q = 0.5$), then the optimal threshold (minimizing total error rate) is at the point where the ROC has slope 1 (i.e., is parallel to the chance diagonal).

2.3.2 The Neyman-Pearson Lemma

Property 3 shows that the slope of the ROC curve at the point with threshold value t is equal to the likelihood ratio

$$\mathcal{L}(t) = \frac{p(t|P)}{p(t|N)}.$$

This ratio tells us how much more probable is a value t of the classifier to have occurred in population P than in population N, which in turn can be interpreted as a measure of confidence in allocation to population P.

However, there is also a much more fundamental connection here with the theory of hypothesis testing. To see this connection, consider the case of testing the simple null hypothesis H_0 that an individual belongs to population N against the simple alternative H_1 that the individual belongs to population P (the hypotheses being simple because the probability models are fully specified for each population). The classification score S is used to make the allocation of the individual, so S forms the data on which the hypothesis test is conducted. Let us suppose that $\mathcal{R}$ is the set of values of S for which we allocate the individual to population P (i.e., the set of values of S for which we reject the null hypothesis). Then the Neyman-Pearson Lemma states that the most powerful test of size α has region $\mathcal{R}$ comprising all values s of S such that

$$\mathcal{L}(s) = \frac{p(s|\text{P})}{p(s|\text{N})} \geq k,$$

where k is determined by the condition $p(s \in \mathcal{R}|\text{N}) = \alpha$. But for the classifier under consideration, $p(s \in \mathcal{R}|\text{N}) = \alpha \Leftrightarrow fp = \alpha$, and the power of the test (i.e., the probability of correctly rejecting the null hypothesis) is just the tp rate.

Thus the connection with the Neyman-Pearson Lemma shows us that for a fixed fp rate, the tp rate will be maximized by a classifier whose set of score values S for allocation to population P is given by $\mathcal{L}(S) \geq k$, where k is determined by the target fp rate. Some implications of this result are that if $\mathcal{L}(S)$ is monotonically increasing then a classification rule based on S exceeding a threshold is an optimal decision rule, and the ROC curve for $\mathcal{L}(S)$ is uniformly above all other ROC curves based on S. For details of these and other properties of $\mathcal{L}(S)$ see Green and Swets (1966), Egan (1975), and Pepe (2003).

2.4 Summary indices of the ROC curve

We have seen above that the ROC is a convenient summary of the full set of information that would be needed for a comprehensive description of the performance of a classifier over all its possible threshold values. However, even such a summary may be too complicated in some circumstances, for instance if a plot is difficult to produce or if very many different classifiers need to be compared, so interest has therefore focussed on deriving simpler summaries. Particular attention has been focussed on single scalar values that might capture the essential features of a ROC curve, motivated by the way that simple summary measures such as mean and variance capture the essential features of statistical data sets. Several such summary indices are now in common use, and we consider the most popular ones.

2.4.1 Area under the ROC curve

Probably the most widely used summary index is the area under the ROC curve, commonly denoted AUC and studied by Green and Swets (1966), Bamber (1975), Hanley and McNeil (1982), and Bradley (1997) among others. Simple geometry establishes the upper and lower bounds of AUC: for the case of perfect separation of P and N, AUC is the area under the upper borders of the ROC (i.e., the area of a square of side 1) so the upper bound is 1.0, while for the case of random allocation AUC is the area under the chance diagonal (i.e., the area of a triangle whose base and height are both equal to 1) so the lower bound is 0.5.

For all other cases, the formal definition is

$$\text{AUC} = \int_0^1 y(x)dx.$$

One immediate interpretation of AUC follows from this definition, elementary calculus and probability theory, plus the fact that the total area of the ROC domain is 1.0: AUC is the average true positive rate, taken uniformly over all possible false positive rates in the range $(0, 1)$. Another interpretation is as a linear transformation of the average misclassification rate, weighted by the mixture distribution of the true P and N classes (see Hand, 2005). It also clearly follows from the above definition that if A and B are two classifiers such that the ROC curve for A nowhere lies below the ROC curve for B, then AUC for A must be greater than or equal to AUC for B. Unfortunately, the reverse implication is not true because of the possibility that the two curves can cross each other.

A less obvious, but frequently used interpretation of AUC for a classifier is that it is the probability that the classifier will allocate a higher score to a randomly chosen individual from population P than it will to a randomly and independently chosen individual from population N.

2.4. SUMMARY INDICES OF THE ROC CURVE

In other words, if S_P and S_N are the scores allocated to randomly and independently chosen individuals from P and N respectively, then

$$\text{AUC} = p(S_P > S_N).$$

To prove this result, we start from the definition of AUC given above and change the variable in the integration from the *fp* rate x to the classifier threshold t. From the proof of property 3 of the ROC as given above we first recollect that

$$y(t) = p(S > t|\text{P}), \quad x(t) = p(S > t|\text{N}), \quad \text{and} \quad \frac{dx}{dt} = -p(t|\text{N}),$$

and we also note that $x \to 0$ as $t \to \infty$ and $x \to 1$ as $t \to -\infty$. Hence

$$\begin{aligned}
\text{AUC} &= \int_0^1 y(x)\,dx \\
&= \int_{+\infty}^{-\infty} y(t)\frac{dx}{dt}\,dt \quad \text{(on changing the variable of integration)} \\
&= -\int_{+\infty}^{-\infty} p(S > t|\text{P})p(t|\text{N})\,dt \quad \text{(from the result above)} \\
&= \int_{-\infty}^{+\infty} p(S > t|\text{P})p(t|\text{N})\,dt \\
&= \int_{-\infty}^{+\infty} p(S_P > t \,\&\, S_N = t)\,dt \quad \text{(by independence assumption)} \\
&= \int_{-\infty}^{+\infty} p(S_P > S_N|t)\,dt \\
&= p(S_P > S_N) \quad \text{(by total probability theorem)}
\end{aligned}$$

as required.

Another interesting interpretation of AUC arises from its connections with the Lorenz curve and the Gini coefficient. The Lorenz curve has long been used in Economics as a graphical representation of an income distribution (Lorenz, 1905), but it has a much wider usage for comparing the disparity in cumulative distributions of two quantities, X and Y say. The usual way it is expressed for income distributions is by ordering the quantities from the least important to the most important, but for our purposes we order them the other way round. In this case, the Lorenz curve is formed by plotting the top percentage of Y corresponding to each top percentage of X (e.g., the top percentage of total wealth owned by the top x percent of the population for all x).

The straight line at 45 degrees from the origin on this plot is the "line of perfect equality" (the top x percent of the population own the top x percent of the wealth for all x), while curves rising progressively further above this line show progressively greater inequalities in the two distributions (progressively smaller percentages owning progressively greater percentages of wealth). The greater the gap between the curve and the line, the greater the inequality in the two distributions. This led Gini (1912) to propose the area between the Lorenz curve and the line of perfect equality, divided by the area below the latter, as a measure of inequality of distributions. This measure is known as the Gini coefficient.

We see from the above descriptions that the ROC is in effect the Lorenz curve that represents the inequality between the *tp* and *fp* rates of a classifier, and that the chance diagonal of the ROC corresponds to the line of perfect equality of the Lorenz curve. The Gini coefficient g can therefore be used to measure the degree of inequality. But since the area below the chance diagonal is 0.5, it follows that

$$g = \frac{\text{AUC} - 0.5}{0.5},$$

so that $2\text{AUC} = g + 1$.

2.4.2 Single points and partial areas

If a specific *fp* rate x_0 is of interest, then the relevant summary from the ROC curve is the single point $y(x_0)$. This is of course just the *tp* rate corresponding to a *fp* rate of x_0 for the given classifier. However, this summary is not perhaps a very useful one for practical purposes. If just a single threshold is envisaged then it is better to calculate all four rates previously defined, while if several classifiers are being compared then it may be difficult to control the *fp* rate at a specifed value x_0 and hence $y(x_0)$ may not be calculable for all classifiers. Perhaps its main use comes in medical applications, where population P often corresponds to a "disease" group while N is the "normal" group. So if the *fp* rate x_0 is a small value (e.g., 0.05), then $100(1-x_0)\%$ can be taken to represent the upper limit (e.g., 95% limit) of the "normal" range of classifier scores. Then $y(x_0)$ is interpreted as the proportion of the disease group that exhibit a "higher than normal" score for this classifier.

More commonly, interest centers on a range of values (a, b) of *fp* that is greater than a single value but less than the full range $(0, 1)$.

2.4. SUMMARY INDICES OF THE ROC CURVE

In this case a suitable summary index of the ROC is the partial area under the curve,
$$\text{PAUC}(a,b) = \int_a^b y(x)dx.$$
However, while this seems a perfectly reasonable way of restricting attention to a portion of the *fp* range that is of particular interest or relevance, the problem is that both the maximum and minimum values depend on a and b which makes interpretation of any one value, and comparison of two or more values, somewhat problematic. On the other hand, the maximum and minimum values are easy to find and this opens up the possibility of standardizing $\text{PAUC}(a,b)$ values for direct comparison. The maximum M is clearly the area of the rectangle with base $(b-a)$ and height 1 (the full extent of the *tp* range) so $M = (b-a)$; while the minimum m is clearly the area under the chance diagonal between $fp = b$ and $fp = a$, and since the chance diagonal has slope 45 degrees then simple geometry and symmetry establishes that $m = (b-a)(b+a)/2$. Thus, in order to standardize $\text{PAUC}(a,b)$ in such a way as to enable direct comparisons to be made between two or more values, McClish (1989) suggests using the transformed index
$$\frac{1}{2}\left(1 + \frac{\text{PAUC}(a,b) - m}{M - m}\right)$$
which has values between 0.5 and 1.0.

As a final comment, we can note that AUC, PAUC, and single-point rates are all special cases of the generalized family of summary values proposed by Weiand et al. (1989). These authors took for their starting point the definition of AUC and PAUC as averages of the true positive rate, taken uniformly over false positive rates in appropriate intervals. To provide a generalization, they then introduced a weighting distribution for the false positive rates and used the resulting weighted average of the true positive rate to summarize the ROC curve. AUC is thus the special case when the weighting distribution is uniform on $(0, 1)$, PAUC when it is uniform on the specified range (a, b) and zero outside it, and a single-point value when it is a degenerate point-distribution (i.e., a dirac delta function).

2.4.3 Other summary indices

While the two summary indices AUC and $\text{PAUC}(a,b)$ are by far the most popular ones used in practice, many others have also been proposed over the years. For discussion of a selection of such indices see Lee and Hsiao (1996) or Lee (1999). We simply note here the equivalence between two of the more common simple indices, both involving maxima. The first is the *Youden Index* YI, which is the maximum difference between the true positive and false positive rates. Thus $YI = \max(tp - fp) = \max(tp + tn - 1)$ since $tn = 1 - fp$. The threshold t at the point on the ROC curve corresponding to the Youden Index is often taken to be the optimal classification threshold. The second simple index is the maximum vertical distance MVD between the chance diagonal and the ROC curve, i.e., MVD $= \max |y(x) - x|$. By using the parametrizations of x and y in terms of the threshold t, and appealing to their probabilistic definitions, it follows that

$$\text{MVD} = \max_t |y(t) - x(t)| = \max_t |p(S > t|\text{P}) - p(S > t|\text{N})|.$$

So MVD is the maximum distance between the cumulative distributions of S in P and N, and ranges between 0 and 1. However, we see that the second term above is just the maximum of $tp - fp$, so MVD is eqivalent to YI. We shall see its further equivalence to the Kolmogorov-Smirnov statistic in Chapter 4. We may also note from Sections 2.2.4 and 2.3.1 that, if costs of making errors are inversely proportional to the relative proportions of P and N individuals, then YI is equivalent to the minimum cost-weighted error rate C.

As well as developing summary indices from a ROC plot, efforts have also been directed over the years at providing alternative plots and associated measures. One important situation is when some information is available about the relative severity of losses due to misclassification, or in other words the case of differential costs of misclassification mentioned earlier. Adams and Hand (1999) point out two drawbacks of the ROC plot in such situations: an awkward projection operation is required to deduce the overall loss; and AUC essentially assumes that *nothing* is known about the relative costs, which is rarely the case in practice. They therefore propose a different kind of plot, the *loss difference* plot, and a new index, the loss comparison (LC) index. However, this takes us beyond the ROC curve itself, so we will pick up the discussion in the next chapter.

2.5 The binormal model

The normal probability distribution has long formed a cornerstone of statistical theory. It is used as the population model for very many situations where the measurements are quantitative, and hence it underpins most basic inferential procedures for such measurements. The reasons for this are partly because empirical evidence suggests that many measurements taken in practice do actually behave roughly like observations from normal populations, but also partly because mathematical results such as the central limit theorem show that the normal distribution provides a perfectly adequate *approximation* to the true probability distribution of many important statistics. The normal model is thus a "standard" against which any other suggestion is usually measured in common statistical practice.

Likewise, for ROC analysis, it is useful to have such a standard model which can be adopted as a first port of call in the expectation that it will provide a reasonable analysis in many practical situations and against which any specialized analysis can be judged. Such a benchmark is provided by the *binormal model*, in which we assume the scores S of the classifier to have a normal distribution in each of the two populations P and N. This model will always be "correct" if the original measurements $\boldsymbol{X}$ have multivariate normal distributions in the two populations and the classifier is a linear function of the measurements of the type first derived by Fisher (1936), as is shown in any standard multivariate text book (e.g., Krzanowski and Marriott, 1995, pp. 29-30). However, it is also *approximately* correct for a much wider set of measurement populations and classifiers. Moreover, as we shall see later in this section, this class is even wider in the specific case of ROC analysis. First, however, let us explore some of the consequences of the binormal assumption.

To be specific, we assume that the distributions of the scores S are normal in both populations and have means μ_P, μ_N and standard deviations σ_P, σ_N in P and N respectively. In accord with the convention that large values of S are indicative of population P and small ones indicative of population N we further assume that $\mu_P > \mu_N$, but we place no constraints on the standard deviations. Then $(S - \mu_P)/\sigma_P$ has a standard normal distribution in P, and $(S - \mu_N)/\sigma_N$ has a standard normal distribution in N. Suppose that the *fp* rate is x, with corresponding classifier threshold t. Then

$$x(t) = p(S > t | \mathrm{N}) = p(Z > [t - \mu_N]/\sigma_N)$$

where Z has a standard normal distribution. Thus

$x(t) = p(Z \leq [\mu_N - t]/\sigma_N)$ (by symmetry of the normal distribution),

so $x(t) = \Phi\left(\frac{\mu_N - t}{\sigma_N}\right)$ where $\Phi(\cdot)$ is the normal cumulative distribution function (cdf). Thus if z_x is the value of Z giving rise to this cdf, then

$$z_x = \Phi^{-1}[x(t)] = \frac{\mu_N - t}{\sigma_N}$$

and

$$t = \mu_N - \sigma_N \times z_x.$$

Hence the ROC curve at this *fp* rate is

$$y(x) = p(S > t|\mathrm{P}) = p(Z > [t - \mu_P]/\sigma_P) = \Phi\left(\frac{\mu_P - t}{\sigma_P}\right),$$

and on substituting for the value of t from above we obtain

$$y(x) = \Phi\left(\frac{\mu_P - \mu_N + \sigma_N \times z_x}{\sigma_P}\right).$$

Thus the ROC curve is of the form $y(x) = \Phi(a + bz_x)$ or $\Phi^{-1}(y) = a + b\Phi^{-1}(x)$, where

$$a = (\mu_P - \mu_N)/\sigma_P, \text{ and } b = \sigma_N/\sigma_P.$$

It follows from the earlier assumptions that $a > 0$, while b is clearly nonnegative by definition. The former is known as the intercept of the binormal ROC curve, and the latter as its slope.

Figure 2.2 shows three ROC curves derived from simple binormal models. The top curve (dotted line) is for the case $\mu_N = 0, \mu_P = 4, \sigma_N = \sigma_P = 1$; a mean difference in classification scores of 4 standardized units represents virtually complete separation of the two normal populations, so the ROC curve is very close to the best possible one. The middle curve (solid line) is for $\mu_N = 0, \mu_P = 2, \sigma_N = \sigma_P = 1$, and the reduction in separation of means to 2 standardized units is reflected in the poorer ROC curve. Finally, the bottom curve (dashed line) has the same values as the middle one except for $\sigma_P = 2$; the further

2.5. THE BINORMAL MODEL

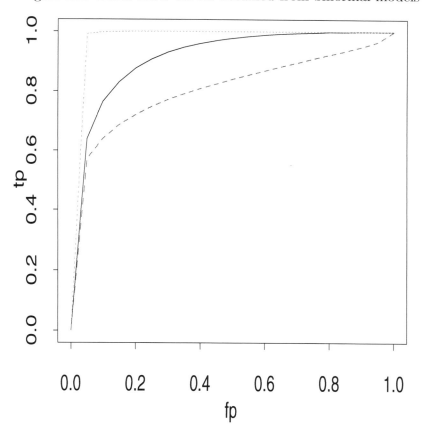

Figure 2.2: Three ROC curves obtained from binormal models

deterioration in performance is caused by the imbalance in standard deviations of the two populations, the higher standard deviation in population P effectively diluting the difference in population means. It might also be mentioned at this point that if one distribution has a sufficiently large standard deviation relative to that of the other, and to the difference between the population means, then the ROC curve will dip below the chance diagonal.

One very useful consequence of this model is that its AUC can be derived very easily, and has a very simple form. We saw earlier that AUC $= p(S_P > S_N) = p(S_P - S_N > 0)$ for independent S_P, S_N. But standard theory tells us that if $S_P \sim N(\mu_P, \sigma_P^2)$ independently of $S_N \sim N(\mu_N, \sigma_N^2)$, then $S_P - S_N \sim N(\mu_P - \mu_N, \sigma_P^2 + \sigma_N^2)$. Hence, if Z denotes a standard normal random variable,

$$\begin{aligned} \text{AUC} &= p\left(Z > 0 - (\mu_P - \mu_N)/\sqrt{\sigma_P^2 + \sigma_N^2}\right) \\ &= 1 - \Phi\left(\frac{-\mu_P + \mu_N}{\sqrt{\sigma_P^2 + \sigma_N^2}}\right) \\ &= \Phi\left(\frac{\mu_P - \mu_N}{\sqrt{\sigma_P^2 + \sigma_N^2}}\right) \\ &= \Phi\left(\frac{a}{\sqrt{1 + b^2}}\right) \end{aligned}$$

on dividing numerator and denominator of the preceding fraction by σ_P. We shall see the utility of this simple form when fitting binormal models to ROC data. Unfortunately, there is no correspondingly simple analytical form for PAUC(a, b), which must therefore be evaluated either by numerical integration or by some approximation formula in any specific application.

Returning to the question of applicability of the binormal model to ROC data, we have derived the above expressions on the assumption that the scores S of the classifier follow normal distributions in each of the two populations. However, we have seen earlier (in Property 2) that the ROC curve is unchanged if a monotone increasing transformation is applied to the scores. Now, although some monotone transformations preserve normality of the populations, others do not. So it is evident that the binormal model will be appropriate for such populations also. Taking the converse viewpoint, the binormal model will be appropriate for any ROC curve pertaining to populations that can be transformed to normality by *some* monotone transformation. This is quite a weak assumption, and hence the binormal model can be applicable to data that do not themselves look particularly normal (see, e.g., Hanley, 1996). It is therefore a very robust model that has found widespread use in ROC analysis

2.6 Further reading

For an introductory article on ROC analysis, see the excellent overview by Fawcett (2006) in an issue of the journal *Pattern Recognition Letters* devoted to ROC analysis in pattern recognition. Although there is no single book devoted exclusively to ROC curves, the topic is central to a number of diverse areas and so constitutes important parts of books in these areas. Further insights into the material covered in this chapter can therefore be gained by reading the appropriate sections in the books by Egan (1975, on signal detection theory), Hand (1997, on classification rules in general), Zhou *et al.* (2002, on diagnostic medicine), and Pepe (2003, on medical tests).

Chapter 3

Estimation

3.1 Introduction

Chapter 2 was concerned exclusively with theoretical concepts in the development of ROC analysis. Thus we merely assumed the existence of a random vector $\boldsymbol{X}$ of measurements taken on individuals, each of which belonged to one of two populations P and N, together with a classification function $S(\boldsymbol{X})$ that could be used to allocate any chosen individual to one of these populations. The mechanism for doing so was to relate the classification score $s(\boldsymbol{x})$ of an individual having value $\boldsymbol{x}$ of $\boldsymbol{X}$ to a threshold t, and to allocate this individual to P or N according as the score exceeded or did not exceed t. Furthermore, if $p(s|\text{P})$ and $p(s|\text{N})$ denote the probability density functions of the classification scores pertaining to individuals coming from populations P and N, then the four fundamental quantities $tp = p(s > t|\text{P})$, $fp = p(s > t|\text{N})$, $tn = p(s \leq t|\text{N})$, and $fn = p(s \leq t|\text{P})$ governed much of the rest of the development.

In practice, of course, we hardly ever know anything for sure about these underlying quantities or populations, but have to work with empirically gathered data sets. Thus, for example, a clinical researcher investigating how to distinguish between those jaundice patients who need surgery and those who can be treated medically, on the basis of the series of measurements $\boldsymbol{X}$ taken on each patient, may know very little about the characteristics of either the "surgical" or the "medical" populations of patients. All that the researcher will typically have available for any decision making will be the values of $\boldsymbol{X}$ for a set of patients known (from either successful treatment or subsequent post-mortem examination) to have come from the "surgical" population, and further values of X for a set of patients from the "medical" population. These sets of values are samples from the relevant populations, so the researcher will need to use the sample values to deduce the relevant population quantities, a process known as *statistical inference*.

The first stage of such inference, on which we concentrate in this chapter, is that of *estimating* the various quantities of interest. This stage may reach right back to the estimation of an appropriate classification rule and its various error and success rates. Strictly speaking this aspect is outside the remit of the present book, but we make some comments about it in Section 3.2. Of more direct relevance to us is the estimation of the ROC curve, and this can be tackled in a number of ways. The most straightforward is the empirical approach, in which no assumptions are made other than that the data form samples from the relevant populations. The ROC curve is then built up directly from data-based estimates of the required true positive and false positive rates. This is essentially a simple method that is easy to implement, and one that has some useful features, but the resulting ROC curve is much more jagged in appearance than its assumed underlying continuous form. Considerable efforts have therefore been directed towards finding estimates that are also smooth continuous functions. One possible line of attack is via parametric modeling of the two populations. Different models yield different functional forms of the ROC curve, so estimation of model parameters leads to smooth ROC curve estimates. If there are concerns that distributional assumptions may not be met, then it is possible to model the ROC curve itself as a smooth parametric function. The binormal model leads to an appropriate such function, so various approaches have therefore been suggested for fitting a binormal model to the data. Although this model assumes that the classification scores can be transformed to normality, the transformation does not have to be explicitly defined and the properties of the model as shown in Chapter 2 imply that it is applicable fairly generally. On the other hand, if it is felt that this model is too restrictive then a nonparametric approach can be adopted, in which no distributional assumptions are made at all and such data-based techniques as kernels and splines are used to produce smoothed ROC curves. Finally, there is a collection of methods involving modeling of probabilities instead of distributions that can be adapted to the case of ROC curves. Each of these approaches is considered in turn in Section 3.3.

Of course, whenever statistical estimation is undertaken it is important to have some idea of the accuracy and precision that can be ascribed to the various estimates, and ROC analysis is no exception in this respect. Accordingly, we also need to consider the sampling variability of the estimators so that their precision can be assessed and so that confidence intervals can be constructed for the relevant ROC features. Also, we need to pay regard to such concepts as bias and mean squared error of the various estimators in order to verify their accuracy. We therefore provide all the necessary results in Section 3.4; they will also be required later, when testing hypotheses about a single curve in Chapter 4 and comparing several curves in Chapter 6.

Finally, having estimated the ROC curve itself, we then need to consider the estimation of the various summary indices of the curve that were derived in Chapter 2, and indeed to consider whether any preferable indices exist for data-based situations. This is covered in Section 3.5.

3.2 Preliminaries: classification rule and error rates

Chapter 2 was built on the assumption that we had available a classifier that would deliver a score $s(\boldsymbol{x})$ when applied to an individual having values $\boldsymbol{x}$, and that these scores were orientated in such a way that larger values were indicative of population P, smaller ones of population N; but we did not discuss the genesis of the classifier. Likewise, all the results on ROC analysis in the data-based case rest on a similar assumption, that a classifier is available to deliver classification scores having the same orientation as above. Thus the researcher who does not yet have such a classifier will first need to select one and build it from the available data before embarking on ROC analysis. Moreover, estimation of the various error and success rates is integral to the data-based process, so we first briefly review these aspects.

There is now an almost embarrassingly wide choice of classifier types: functions such as the linear or quadratic discriminators that are derived from theoretical models for the data; general-purpose statistical and computational procedures such as logistic discrimination or classification trees; and "black box" routines such as neural networks or support vector machines that generate classifiers. All these methods contain either parameters that have to be estimated or computational choices that have to be made in order to tailor them to the problem in hand, so need input from data before they are fully specified. Moreover, once they are fully specified they either need extra data, or clever ways of re-using the same data, in order to assess their performance via success or error rates. There has been very much work done on all these aspects over the years, and full discussion is clearly outside the scope of this book. Some very general comments have already been made in Chapter 1, and we here fill these out with a brief outline of a few of the most common scenarios. However, the interested reader is referred for further details to specialist texts such as those by McLachlan (1992), Hand (1997), Hastie *et al.* (2001), Webb (2002), or Kuncheva (2004).

If the number of individuals represented in the available data is large, then the favored approach is to divide the set of data randomly into two portions, making sure that there is adequate representation of each group (class, population) in each of the two portions. The first portion is termed the *training set* and is used to either estimate the parameters or to make the computational choices needed to build the classifier, while the second portion is termed the *test set* and is used to derive the various required rates (tp, fp, tn, fn). This is done simply by applying the classifier to each individual in the test set and hence obtaining the set of scores for this set. For a given threshold t each score is converted into a classification, so we can note whether each classification is correct or incorrect and thus estimate the rates by the appropriate ratios of either correct or incorrect classifications.

If the available data set is not large, however, then all of it perforce becomes the "training" set. However, as has already been stressed in Chapter 1, simple reuse of this set for estimating the various rates must be avoided because the resultant estimates will be optimistically biased. Hence one of a number of the more ingenious data-based estimates has to be used for deriving the rates, in order to avoid this bias. Given the need for imposing a range of different threshold values on each classification score, the best of these estimates is arguably the one given by *cross-validation*. This requires the set of individuals to be randomly partitioned into a number of subsets, each one of which is treated in turn as a "test" set. The classifier is built from the rest of the data, and applied to the individuals in the "test" subset to generate their scores. Once all subsets have taken their turn as the "test" set, scores exist for all individuals in the original data and estimation of rates proceeds as before. Of course this process involves considerable computation, as the classifier has to be built anew for each test subset. Also, a decision must be made as to how many subsets there should be. A common choice is about 10, but also often favored is the extreme choice of having a separate sub-group for each individual in the data; this is known as *leave-one-out* cross-validation.

Further complications may arise with some classifiers. For example, *tuning* may be needed during the course of classifier construction. This occurs when monitoring of error rates is necessary to decide when the training process is complete (e.g., in building a neural network, or in deciding when to stop a variable selection process). To accomplish such tuning it is necessary to split the available data into a "training" and a "tuning" set, so that the error rates being monitored are independent of the classifier construction. Therefore, since we additionally need a set from which to estimate the rates for the ROC, we either have to split the available data randomly into *three* sets at the outset, or we need to conduct the cross-validation on the training set and leave the tuning set untouched throughout the process, or we need to perform a *nested* cross-validation on the complete data in order to accommodate both the tuning and the rate estimation. The important point is that for the ROC analysis to be reliable we need to have a set of scores from which the various required rates have been calculated *independently* of any calculations done during the building of the classifier.

3.3 Estimation of ROC curves

Since we are dealing exclusively with continuous scores S, recollect from 2.2.4 that in this case the ROC curve has equation

$$y = 1 - G[F^{-1}(1-x)] \quad (0 \leq x \leq 1),$$

where f, F are the density and distribution functions of S in population N and g, G are the density and distribution functions of S in population P. The estimation problem thus reduces to the problem of estimating this curve from the given data.

3.3.1 Empirical estimator

To obtain the empirical estimator, we simply apply the relevant definitions from Chapter 2 to the observed data. Thus, if n_P and n_N are the numbers of individuals in the samples from populations P and

N respectively, and if $n_{A(t)}$ denotes the number of individuals in the sample from population A (where A is either N or P) whose classification scores are greater than t, then the empirical estimators of the true positive rate $tp = p(S > t|P)$ and false positive rate $fp = p(S > t|N)$ at the classifier threshold t are given by

$$\widehat{tp} = \frac{n_{P(t)}}{n_P},$$

and

$$\widehat{fp} = \frac{n_{N(t)}}{n_N}.$$

Thus plotting the set of values $1 - \widehat{fp}$ against t yields the empirical distribution function $\hat{F}(t)$, and doing the same for values $1 - \widehat{tp}$ yields the empirical distribution function $\hat{G}(t)$.

The empirical ROC curve is then simply given by plotting the points $(\widehat{fp}, \widehat{tp})$ obtained on varying t, so is the curve

$$y = 1 - \hat{G}[\hat{F}^{-1}(1-x)] \quad (0 \leq x \leq 1).$$

Although technically all possible values of t need to be considered, in practice $\widehat{fp}$ will only change when t crosses the score values of the n_N individuals and $\widehat{tp}$ will only change when t crosses the score values of the n_P individuals, so there will at most be $n_N + n_P + 1$ discrete points on the plot. Moreover, if there are ties in score values in either group or between groups this will reduce the number of distinct points, and it can be verified easily that the actual number of points will be one more than the number of distinct score values across both groups. The points are joined by lines, and this produces the jagged appearance alluded to earlier: the lines are either horizontal or vertical if just one of $(\widehat{fp}, \widehat{tp})$ changes at that value of t, and they are sloped if both estimates change.

To show how the process works, consider a small fictitious set of illustrative data with 10 individuals in each group, such that the scores produced by a classifier for these individuals are

0.3, 0.4, 0.5, 0.5, 0.5, 0.6, 0.7, 0.7, 0.8, 0.9 for group N and
0.5, 0.6, 0.6, 0.8, 0.9, 0.9, 0.9, 1.0, 1.2, 1.4 for group P.

Table 3.1 shows the calculations for each distinct value of t across all the data (noting that the origin of the plot occurs at the *highest* value of t). To see how the values in the table arise, consider progressively

3.3. ESTIMATION OF ROC CURVES

lowering the value of t. For any value greater than or equal to 1.4, all 20 observations are allocated to group N, so no P individuals are allocated to P (hence $\widehat{tp} = 0.0$) and all the N individuals are allocated to N (hence $\widehat{fp} = 0.0$). On crossing 1.4 and moving t down to 1.2, one individual (the last) in group P is now allocated to P (hence $\widehat{tp} = 0.1$) while all the N individuals are still allocated to N (hence $\widehat{fp} = 0.0$ again). Continuing in this fashion generates all the results in the table, and Figure 3.1 shows the plotted ROC curve together with the chance diagonal.

Table 3.1: Points in the empirical ROC curve of the fictitious data.

t	$\widehat{fp}$	$\widehat{tp}$
≥ 1.4	0.0	0.0
1.2	0.0	0.1
1.0	0.0	0.2
0.9	0.0	0.3
0.8	0.1	0.6
0.7	0.2	0.7
0.6	0.4	0.7
0.5	0.5	0.9
0.4	0.8	1.0
0.3	0.9	1.0
<0.3	1.0	1.0

Some general features of such empirical curves are worth noting. They are increasing step functions, with typical steps of size $\frac{1}{n_N}$ horizontally and $\frac{1}{n_P}$ vertically. However, if there are tied scores in one of the groups then the corresponding step will be larger in the appropriate direction, with size depending on number of tied values. Thus the typical step size is 0.1 in both directions for the fictitious data, but the two score values 0.7 for the data from group N lead to the longer horizontal step of size 0.2 between 0.2 and 0.4. Also, if there are ties between the scores across both groups, the lines on the plot will be diagonal rather than horizontal or vertical, as for example between (0.0, 0.3) and (0.1, 0.6) (one 0.9 in N and three in P), between (0.4, 0.7) and (0.5, 0.9) (one 0.6 in N and two in P), or between (0.5, 0.9) and (0.8, 1.0) (three 0.5s in N and one in P).

It can also be noted that the empirical ROC curve depends only

Figure 3.1: Empirical ROC curve, plus chance diagonal for the fictitious data

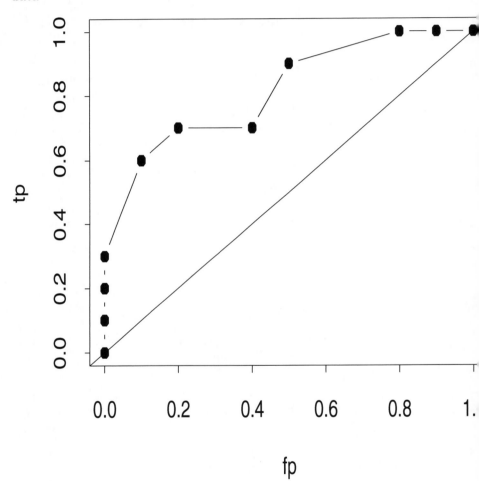

3.3. ESTIMATION OF ROC CURVES

on the ranks of the combined set of test scores (Zweig and Campbell, 1993). This has the advantage of making it invariant with respect to monotone transformation of test scores. Thus if during the course of a practical analysis it becomes preferable to work with, say, logarithms rather than original values, then no recalculations of the curve are necessary. However, Fawcett (2006) shows through the following simple example that misleading inferences can be made from the empirical ROC curve about the accuracy of a classifier. For a set of ten scores obtained from a Naive Bayes classifier, the constructed ROC curve rises vertically from (0, 0) to (0, 1) and is then horizontal to (1, 1), which suggests an optimum classifier. The analyst might therefore be tempted to infer that it will yield a perfect classificatory performance. However, assuming that the scores are on a probability scale and using the threshold $t = 0.5$ on them produces two misclassifications and hence a classifier accuracy of only 80%. The discrepancy arises because the ROC curve merely shows the ability of the classifier to correctly rank the P and N scores, while the accuracy measure results from the imposition of an inappropriate threshold. The accuracy would have been 100% had the scores been correctly calibrated, or had the threshold been changed to $t = 0.7$. The example therefore shows that care must be taken when attempting to draw conclusions from a curve. Nevertheless, computation of empirical ROC curves is straightforward and programs or algorithms are documented in the literature (e.g., Vida, 1993; Fawcett, 2006).

To provide some larger real examples, consider two of the data sets obtained from the UCI repository of machine learning databases (Mertz and Murphy, 1996). The first set consists of observations taken on 768 individuals belonging to the Pima tribe of Indians, 268 of whom tested positive for diabetes (group P) and 500 of whom tested negative (group N). The values of eight variables were recorded for each individual, and the data were randomly divided into a training set of 538 and a test set of 230. A quadratic discriminant function and a neural network were both built on the training data, and Figure 3.2 shows the empirical ROC curves resulting from each classifier.

The second example comes from the heart disease directory of the UCI repository, and concerns data collected at the Hungarian Institute of Cardiology in Budapest by Andros Janosi, MD. Measurements on 14 variables were taken from 98 patients suffering from heart disease (group P) and 163 subjects who did not have heart disease (group N),

Figure 3.2: Empirical ROC curves for quadratic classifier and neural network on the Pima Indian data

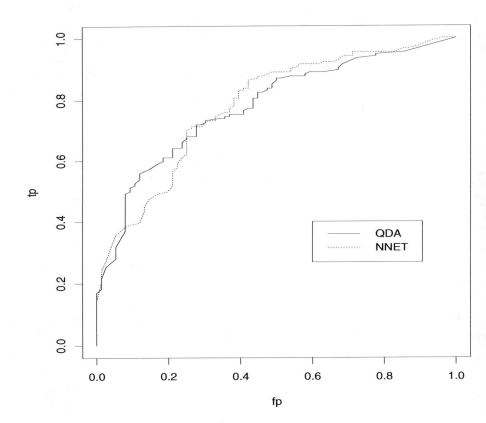

3.3. ESTIMATION OF ROC CURVES

and the data were randomly divided into a training set of 183 subjects and a test set of 78 subjects. Linear and quadratic discriminant analyses were both conducted on the data, and Figure 3.3 shows the two resulting empirical ROC curves.

3.3.2 Parametric curve fitting

If the irregular appearance of the empirical ROC curve is not deemed to be adequate as an estimate of the underlying "true" smooth curve, then one of the alternative methods of estimation must be adopted. The obvious approach from a traditional statistical perspective is to assume parametric models for the distributions of the classification statistic in both populations, which in turn provides induced smooth parametric forms for both distribution functions $F(x)$ and $G(x)$. Estimation of model parameters using maximum likelihood on the sample data, and substitution of parameters in $F(x)$ and $G(x)$ by these estimates, then yields smooth estimates of the ROC curve

$$y = 1 - G[F^{-1}(1-x)] \quad (0 \leq x \leq 1).$$

To plot the curve, of course, we do not need to carry out any complicated function inversion, but simply need (from the definition of the ROC curve) to plot $1 - G(x)$ against $1 - F(x)$ for a range of values of x in $(0, 1)$. Zweig and Campbell (1993) surveyed possible parametric models that can be used with this approach. A normal model in each population is perhaps the obvious choice, but logistic (Ogilvie and Creelman, 1968), Lomax (Campbell and Ratnaparkhi, 1993), and gamma (Dorfman et al. 1997) models have been among other ones considered. However, this approach relies critically on the distributional assumptions being appropriate, and if they are violated in any way then the consequences can be serious—if not on the curve itself, then on derived quantities such as AUC (see, e.g., Goddard and Hinberg, 1990). Hence Zhou et al. (2002) caution against using this approach unless the data have been checked carefully for consistency with the assumptions. In this regard, Zou and Hall (2000) propose both parametric and semiparametric transformation models that make the approach more defensible. The parametric method is the more straightforward. This requires a Box-Cox transformation of the classification statistic to normality, with all parameters being estimated by maximum likelihood, followed by construction of the ROC curve from the transformed values.

Figure 3.3: Empirical ROC curves for linear and quadratic discriminant functions applied to the Budapest heart data

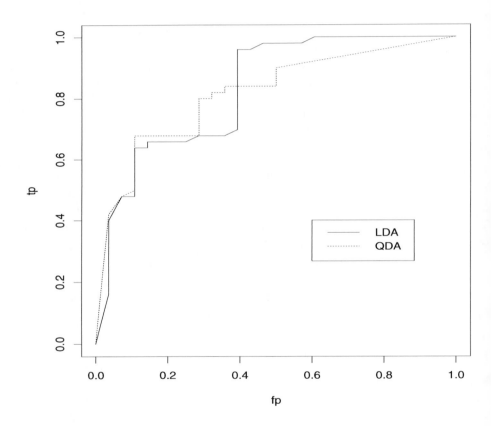

3.3. ESTIMATION OF ROC CURVES

However, the binormal model discussed in Section 2.5 is a more flexible distributional approach. It only assumes that the data can be transformed to normality by means of some monotonic transformation, and this transformation itself need not be specified. Moreover, we have seen in Chapter 2 that the population ROC curve is unchanged under arbitrary monotonic transformation of the classification statistic, so this property will extend to estimates of the curve based on the binormal model. Unfortunately, curves estimated on the basis of specific distributional assumptions about each separate population are generally unlikely to satisfy this property. Consequently, much more attention has been given to the binormal model than to the other possibilities mentioned above.

We know from Section 2.5 that for the binormal model the ROC curve has form

$$y(x) = \Phi(a + bz_x) \quad (0 \le x \le 1),$$

where

$$a = (\mu_P - \mu_N)/\sigma_P, \quad b = \sigma_N/\sigma_P, \quad z_x = \Phi^{-1}(x)$$

and $\mu_N, \mu_P, \sigma_N, \sigma_P$ are the means and standard deviations of the monotonically transformed classification scores in the two populations. Hence the problem reduces to the estimation of parameters a and b. A graphical check on the goodness-of-fit of the binormal model may be obtained by plotting the empirical ROC curve using normal probability scales on both axes, where a roughly linear plot of the points indicates that the model is appropriate. Moreover, if it is deemed to be appropriate, then fitting a straight line to these points yields graphical estimates of the parameters (Swets and Pickett, 1982, page 30) and such estimates would provide good initial values in an iterative refinement process.

The earliest analytical approach is due to Dorfman and Alf (1969), and versions of this approach are still in common use at the present time. It was originally conceived for use on ordered categorical data, where the classification statistic S can take on only one of a finite set of (ranked) values or categories $C_1, C_2, \ldots, C_k$ say. The Dorfman and Alf procedure assumes that there is a latent random variable W, and a set of unknown thresholds $-\infty = w_0 < w_1 < w_2, \ldots < w_k = \infty$, such that S falls in category C_i if and only if $w_{i-1} < W \le w_i$. Identifying the latent variable W as the result of the unknown transformation of the classification statistic to normality thus establishes the binormal model for the data. Hence it follows that

$$p_{iN} = \Pr\{S \in C_i | N\} = \Phi\left(\frac{w_i - \mu_N}{\sigma_N}\right) - \Phi\left(\frac{w_{i-1} - \mu_N}{\sigma_N}\right)$$

and

$$p_{iP} = \Pr\{S \in C_i | P\} = \Phi\left(\frac{w_i - \mu_P}{\sigma_P}\right) - \Phi\left(\frac{w_{i-1} - \mu_P}{\sigma_P}\right).$$

Then if n_{iN}, n_{iP} are the observed numbers of individuals from populations N and P respectively falling in category C_i (for $1 \leq i \leq k$), the log likelihood of the sample is given by

$$\mathcal{L} = \sum_{i=1}^{k}(n_{iN} \log p_{iN} + n_{iP} \log p_{iP}).$$

Dorfman and Alf (1969) developed the iterative algorithm for maximizing this log likelihood with respect to the model parameters in the presence of categorical data, and it still remains the most popular method for such data. Note that Ogilvie and Creelman (1968) developed the same algorithm when fitting the logistic model.

When moving on to continuous data, Metz et al. (1998) noted that one possible approach was to categorize the scores into a finite number of categories, which by definition will be ordered ones, and then to apply the Dorfman and Alf procedure. They described this approach, and implemented it in the computer package LABROC. Subsequent investigations revolved around the question of how dependent the procedure was on the categorization scheme, and whether the categories had to be pre-specified or whether they could be data-dependent, resulting in several updates of the LABROC package (all of which are now publicly available on the World Wide Web). As an illustration, Figure 3.4 shows the result of applying the LABROC algorithm to the quadratic discriminant scores for the Pima Indian data. Comparison with the empirical ROC curve for this classifier in Figure 3.3 demonstrates how much smoother the curve becomes when the binormal assumption is made.

Hsieh and Turnbull (1996) conducted a comprehensive study of parameter estimation for the binormal model in the presence of grouped continuous data, noting the use of the Dorfman and Alf algorithm for such data. However, they commented on the difficulty of the computation when the number k of categories is large, due to the presence of

3.3. ESTIMATION OF ROC CURVES

Figure 3.4: ROC curve for the quadratic classifier on the Pima Indian data, with binormal assumption and using the LABROC algorithm

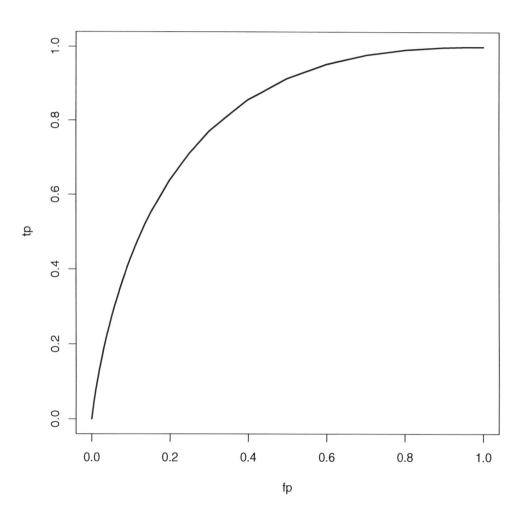

many nuisance parameters $w_1, w_2, \ldots, w_k$, which sometimes causes the iterative estimation procedure to fail. They therefore proposed two alternative approaches. One was an approximation to the Dorfman and Alf procedure, after noting that the latter can be expressed approximately as a regression model with measurement error. The nuisance parameters can be removed by using sample estimates of p_{iN} and p_{iP}, and this produces a simpler method that is less prone to failure. The second one was a generalized least squares approach, based on the fact that a natural estimator of

$$\beta = \Phi(a + bz_x) \quad (0 \leq x \leq 1)$$

is the empirical ROC value

$$\hat{\beta} = 1 - \hat{G}[\hat{F}^{-1}(1-x)] \quad (0 \leq x \leq 1).$$

Thus if β_i is the value at x_i $(i = 1, \ldots, k)$ then one can write

$$\Phi^{-1}(\hat{\beta}_i) = a + bz_{x_i} + \epsilon_i, \quad (i = 1, \ldots, k).$$

Hsieh and Turnbull (1996) derived the asymptotic covariance structure Σ of the ϵ_i, and hence provided the generalized least squares estimators of a and b. The iterative procedure comprises a first iteration in which the ordinary least squares estimators of a and b are obtained from the regression equation, and further iterations in which Σ is estimated at the current a, b values and generalized least squares is used. Hsieh and Turnbull also gave an adaptive method for finding initial values, and commented that more than two iterations of the process would rarely be needed.

Moving on to general continuous data, Hsieh and Turnbull foreshadowed Metz *et al.* (1998) by pointing out that the data must be grouped in order to use the Dorfman and Alf procedure (or indeed either of the two others above), and this will inevitably lead to some loss of efficiency. They therefore proposed a minimum distance estimator (Wolfowitz, 1957) of the ROC curve for general continuous data, which is in effect the closest curve of binormal form to the empirical ROC curve

$$y(x) = 1 - \hat{G}[\hat{F}^{-1}(1-x)] \quad (0 \leq x \leq 1)$$

defined earlier. If we denote the discrepancy between the empirical ROC curve and the binormal model at the point $X = x$ by $d(x)$, then

$$d(x) = [1 - \hat{G}[\hat{F}^{-1}(1-x) - \Phi(a + bz_x)].$$

3.3. ESTIMATION OF ROC CURVES

The total squared discrepancy between the two curves is then given by

$$d^2 = \int_0^1 d^2(x)dx,$$

and the minimum distance estimators of a, b are the values that minimize d^2. Cai and Moskowitz (2004) focussed on maximum likelihood estimation of the binormal model for ungrouped continuous data. They summarized the problems encountered with this approach, and proposed a maximum profile likelihood estimator together with an algorithm for its numerical implementation. They also derived a pseudo maximum likelihood estimator that can be used as an alternative, and one that is extended easily to include covariate information.

As a final comment in this section, it is worth noting that postulating specific probability models for populations N and P is generally termed a *fully* parametric approach to ROC curve estimation, while assuming that some unknown transformation converts both populations to a specific form (such as the normal form) is often referred to either as a *semiparametric* or a *parametric distribution-free* approach. Also, Hsieh and Turnbull (1996) point out that the binormal model is an example of a two-sample transformation model, cite other possible examples such as logistic and Weibull models, and give further references to work on estimation of such models.

3.3.3 Nonparametric estimation

Approaches described in the previous section share some potential weaknesses, principally that the assumed model or form may be inappropriate for the data at hand, and that the number of possible models is necessarily limited so may not include an appropriate one. Consequently, several authors have turned more recently to nonparametric methods, which are applicable very generally. The idea is to obtain smooth estimates of the functions $F(x)$ and $G(x)$ directly from the data, without imposing any distributional constraints, and then either to plot $1 - \hat{G}(x)$ against $1 - \hat{F}(x)$ for a graphical depiction of the ROC curve, or to obtain an explicit equation for it by substituting for $F(.)$ and $G(.)$ in the defining equation

$$y = 1 - G[F^{-1}(1-x)] \quad (0 \le x \le 1).$$

To discuss such data-based estimation we require notation for individual classification statistic values, so let us suppose that the n_N observations from population N are denoted $s_{N1}, s_{N2}, \ldots, s_{Nn_N}$ while the n_P observations from population P are denoted $s_{P1}, s_{P2}, \ldots, s_{Pn_P}$.

The first approach of this kind was by Zou et al. (1997), who suggested using kernel density methods (Silverman, 1986) for estimating the density functions in each population, followed by integration of the density functions to obtain estimates of $F(x)$ and $G(x)$. Specifically, kernel estimates of the two density functions are given by

$$\hat{f}(x) = \frac{1}{n_N h_N} \sum_{i=1}^{n_N} k\left(\frac{x - s_{Ni}}{h_N}\right)$$

and

$$\hat{g}(x) = \frac{1}{n_P h_P} \sum_{i=1}^{n_P} k\left(\frac{x - s_{Pi}}{h_P}\right),$$

where $k(\cdot)$ is the kernel function in both populations and h_N, h_P are the bandwidths in each. Received wisdom in kernel density estimation is that choosing between the many available kernel functions is relatively unimportant as all give comparable results, but more care needs to be taken over the selection of bandwidth. Zou et al. (1997) in fact suggest using the biweight kernel

$$k\left(\frac{x-a}{b}\right) = \frac{15}{16}\left[1 - \left(\frac{x-a}{b}\right)^2\right]^2 \quad x \in (a-b, a+b)$$

and $k = 0$ otherwise, with the general-purpose bandwidths (Silverman, 1986) given by $h_N = 0.9\min(sd_N, iqr_N/1.34)/(n_N)^{\frac{1}{5}}$ and $h_P = 0.9\min(sd_P, iqr_P/1.34)/(n_P)^{\frac{1}{5}}$, where sd, iqr denote the sample standard deviation and inter-quartile range respectively for the subscripted populations.

Having obtained $\hat{f}(x)$ and $\hat{g}(x)$ in this way, we then calculate $\hat{F}(t) = \int_{-\infty}^{t} \hat{f}(x)dx$ and $\hat{G}(t) = \int_{-\infty}^{t} \hat{g}(x)dx$ by numerical integration for a set of values of t, and the estimated ROC curve follows directly by plotting $1 - \hat{G}(t)$ against $1 - \hat{F}(t)$. Figure 3.5 shows the result of applying this approach to the quadratic discriminant scores for the Pima Indian data. This nonparametric ROC curve is intermediate in smoothness between the empirical curve of Figure 3.3 and the LABROC curve of Figure 3.4.

Zou et al. (1997) recommend first transforming the classification scores to symmetry if they show evidence of skewness. However, their choice of bandwidth is not optimal for the ROC curve, because the latter depends on the distribution functions $F(x), G(x)$ and optimality

3.3. ESTIMATION OF ROC CURVES

Figure 3.5: Nonparametric kernel-based ROC curve for the quadratic classifier on the Pima Indian data

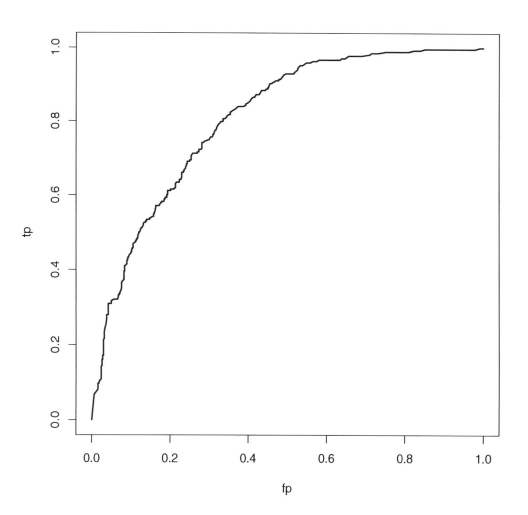

for estimating density functions does not imply optimality for estimating distribution functions. Lloyd (1998) therefore considered the direct estimation of $F(x), G(x)$ using kernel density methods, while Zhou and Harezlak (2002) investigated several different bandwidths. These two studies improved on Zou et al.'s choice, by obtaining asymptotically optimal estimates of the distribution functions $F(x)$ and $G(x)$.

Nevertheless, the resulting estimator still has some drawbacks as pointed out by Peng and Zhou (2004): (i) while the estimates of $F(x)$ and $G(x)$ may be asymptotically optimal, there is no guarantee of asymptotic optimality for the ROC curve itself; (ii) since $F(x)$ and $G(x)$ are estimated separately, the final ROC curve estimator is not invariant under a monotone transformation of the data; and (iii) the boundary problem in smoothing means that the estimator is unreliable at the ends of the ROC curve. Both Peng and Zhang (2004) and Ren et al. (2004) attempted to overcome these drawbacks, and both attacked the problem by smoothing the empirical ROC curve. The former did so by local linear smoothing of the empirical curve values using a kernel smoother; the latter by means of penalized regression splines applied to a linear mixed model. These approaches seem to afford the greatest flexibility while fully utilizing the data.

3.3.4 Binary regression methods

Several authors have recently investigated a less direct way of estimating $F(t) = p(S > t|\text{N})$ and $G(t) = p(S > t|\text{P})$, by reversing the roles of the classification statistic and the population labels and first seeking estimates of $p(\text{N}|S = s)$ and $p(\text{P}|S = s)$. In order to bring these two probabilities within a single formula, we can define a "population" variable by $\mathcal{G}$ having values 0 for population N and 1 for population P, and write $\pi_j(s) = p(\mathcal{G} = j|S = s)$ for $j = 0, 1$. Additionally, denote the overall probability that a sample individual comes from $\mathcal{G} = j$ by π_j for $j = 0, 1$.

Lloyd (2002) approached the problem as follows. First, rewrite the two sets of sample values in the single form as pairs (u_i, s_i) for $i = 1, 2, \ldots, n = n_N + n_P$, where $u_i = 0$ or 1 according as the ith observation is from population N or P, and s_i is the classification statistic value for the ith individual. Then the joint likelihood $\mathcal{L}$ of the n pairs can be factorized:

$$\mathcal{L} = L[(u_1, s_1), \ldots, (u_n, s_n)]$$

$$= L(u_1,\ldots,u_n|s_1,\ldots,s_n)L(s_1,\ldots,s_n)$$
$$= \left(\prod_{i=1}^{n} \pi_0(s_i)^{1-u_i}\pi_1(s_i)^{u_i}\right)\left(\prod_{i=1}^{n}[\pi_0 f(s_i)+\pi_1 g(s_i)]\right).$$

Lloyd (2002) proposed using a binary (i.e., logistic) regression to estimate the $\pi_j(s)$ in terms of any set of chosen basis functions of s. He outlined the steps of the process for maximizing $\mathcal{L}$ in terms of the parameters of the logistic regression function, established that the maximum likelihood estimates of $F(x)$ and $G(x)$ were both cumulative distribution functions of discrete distributions concentrated at the distinct values of $s_1, s_2, \ldots, s_n$, and gave the associated probabilities of both distributions. The estimated $\hat{F}(x), \hat{G}(x)$ can thus be obtained from these probabilities by partial summation.

Qin and Zhang (2003) likewise used a logistic model for the $\pi_j(s)$, viz

$$\pi_j(s) = \frac{\exp[\alpha^* + \beta^T r(s)]}{1 + \exp[\alpha^* + \beta^T r(s)]},$$

where $r(s)$ is a vector of basis functions. Having shown in a previous article that

$$\frac{g(s)}{f(s)} = \exp[\alpha + \beta^T r(s)],$$

where $\alpha = \alpha^* + \log(\pi_0/\pi_1)$, this gives a connection between the two density functions. Qin and Zhang then went on from this connection to derive what they termed maximum semiparametric estimators of the ROC curve.

Further methods based on regression and generalized linear modeling have been developed by Pepe and coworkers, but as these methods also allow the incorporation of covariates into the estimation process they are covered in Chapter 5.

3.4 Sampling properties and confidence intervals

Any estimator of a population quantity is essentially a sample statistic (i.e., a function of the n values in the sample) and as such its value will vary from sample to sample. The distribution of all possible values of the estimator under random sampling of sets of n individuals from the population is termed the *sampling distribution* of the estimator.

Knowledge of this distribution, or at least of its main features, provides useful information about the worth of the estimator. The mean of the sampling distribution is the expected value of the estimator under random sampling: if this mean equals the population quantity being estimated then the estimator is said to be *unbiased*, and if it does not equal the population quantity then the difference between the two is the *bias* of the estimator. The bias indicates by how much, on average, the estimator will either exceed (positive bias) or fall short of (negative bias) the population quantity. The standard deviation of the sampling distribution is known as the *standard error* of the estimator, and this quantifies the likely variability in the estimator's values from sample to sample. If the distributional form of the sampling distribution is also known together with its mean and standard deviation, then a *confidence interval* can be constructed for the population quantity. Such an interval gives a range of values within which there is a specified confidence of finding the population quantity, as opposed to a single guess at its value that is almost certain to be wrong.

In many situations, mathematical expressions can be derived for the features of the sampling distribution but these expressions involve population parameters which must be replaced by their estimates from the available sample values. However, when the estimators are complicated functions, as with most ROC curve applications, then often the only viable results are *asymptotic* ones (i.e., valid only for large n). Moreover, in some situations even asymptotic results have proved intractable. In such situations, recourse must be made to one of the data re-sampling methods such as jackknifing or bootstrapping, in order to mimic the drawing of repeated samples and thereby to calculate the necessary quantities in a purely data-based fashion. For further details of these techniques, see Efron and Tibshirani (1993), Davison and Hinkley (1997), or Zhou *et al.* (2002, pages 155-158).

With this brief background, we now summarize the available results for the estimators described earlier.

3.4.1 Empirical estimator

Rigorous mathematical derivation of the asymptotic distribution of the empirical estimator has been provided by Hsieh and Turnbull (1996), but a more practically-orientated account is given by Pepe (2003) so we follow the latter here.

Consider a point $(\widehat{fp}, \widehat{tp})$ on the empirical ROC curve. This esti-

3.4 SAMPLING RESULTS AND INTERVAL ESTIMATES

mates the "true" point (fp, tp), which arises when threshold t is applied to the classification score S. If we ask about the likely variability of the estimated point over repeated sampling from the same population, there are actually three distinct types of variability we could envisage: (vertical) variation in $\widehat{tp}$ over samples having fixed fp; (horizontal) variation in $\widehat{fp}$ over samples having fixed tp; and (two-dimensional) variation in $(\widehat{fp}, \widehat{tp})$ over samples having fixed classification score threshold t. The last version is actually the easiest to handle, because we know from section 3.3.1 that $\widehat{tp} = \frac{n_{P(t)}}{n_P}$ and $\widehat{fp} = \frac{n_{N(t)}}{n_N}$, and the samples from the two populations N and P are assumed to be independent of each other. Thus, providing that individuals in each sample are mutually independent, $\widehat{fp}$ and $\widehat{tp}$ are independent estimates of binomial probabilities so standard exact binomial methods can be applied directly to obtain confidence intervals $(\widehat{fp}_L, \widehat{fp}_U)$ for fp and $(\widehat{tp}_L, \widehat{tp}_U)$ for tp. Since the two samples are independent, the multiplication rule of probabilities establishes that if a $100(1-\alpha)\%$ confidence interval for (fp, tp) is desired, it is given by the rectangle whose sides are the $100(1-\tilde{\alpha})\%$ confidence intervals for fp and tp separately, where $\tilde{\alpha} = 1 - \sqrt{1-\alpha}$. If the individuals in the two samples are not independent, however, then no analytical approach is available and recourse must be made to one of the data-based methods.

Next consider the case of vertical variation in $\widehat{tp}$ over samples having fixed fp. The fixed fp determines the true threshold t. However, we only have an estimate $\widehat{fp}$ of fp, so from this estimate we can obtain the *estimated* $\hat{t}$. This process involves only the n_N observations in the sample from population N. Having obtained $\hat{t}$, we determine the number of n_P observations exceeding it in order to obtain $\widehat{tp}$, and this only involves the observations in the sample from population P. So if we now imagine taking repeated samples from the two populations and looking at the variation in $\widehat{tp}$ values at a fixed fp, there are two sources contributing to this variation: variability in the estimated thresholds $\hat{t}$ (involving sampling variability in observations from N) and variability in $\widehat{tp}$ at a given value of $\hat{t}$ (involving sampling variability in observations from P). Since the two samples are independent, the variance of the point in the vertical direction will thus be the sum of the variances from these two sources. Indeed, Pepe (2003, Chapter 5) gives the full asymptotic distribution in the case when the observations in each sample are mutually independent. To state it most concisely, let us return to the original notation for the ROC curve in the form $y = h(x)$,

where $y = tp$ and $x = fp$, so that $\hat{y} = \widehat{tp}$. Then for large n_N, n_P, the distribution of $\hat{y}$ at fixed value of x is asymptotically normal with mean y and variance
$$\frac{y(1-y)}{n_P} + \left(\frac{g(c)}{f(c)}\right)^2 \frac{x(1-x)}{n_N},$$
where c is determined by $x = p(S > c|N) = 1 - F(c)$.

The two terms in the variance correspond to the two sources of variability identified above: the first term involves the sample data only through n_P so is the variability in the binomial probability estimate, while the second term involves the sample data only through n_N so is the variability in the estimated thresholds. This latter term includes the likelihood ratio $\frac{g(c)}{f(c)}$, which we saw in property 3 of Section 2.2.3 to equal the slope of the true curve $y = h(x)$. So large slopes are associated with high variances of $\hat{y}$ (since small changes in estimated x values are likely to have a large effect). By contrast, at both ends of the ROC curve, where x and y are both close to either 0 or 1, the variance will be small because of the small amount of "slack" available in the estimation process.

In the case of horizontal variation in $\hat{x}$ for a fixed value of y, it is easy to see that all the same arguments follow but with the roles of the two populations N and P interchanged. So we can use the same result as above but interchanging n_N and n_P, $f(c)$ and $g(c)$, and x and y respectively, and determining c from $y = p(S > c|P) = 1 - G(c)$.

Of course, if either of the two variances are to be used in practice to obtain a confidence interval for y (from the first variance) or x (from the second), then unknown quantities must all be replaced by their sample estimates. Thus in the first case x is given (at the fixed value), but y is replaced by $\widehat{tp}$ and c is obtained by finding the smallest value exceeded by a proportion x of the sample from population N. The two density functions f, g must also be estimated at the value c. This can be done either by using a kernel density estimate, as in the nonparametric method due to Zou *et al.* (1997) described above in Section 3.3.3, or much more simply by the number of observations from each sample lying between $c - r$ and $c + r$ divided by $2rn_N$ for $f(c)$ and $2rn_P$ for $g(c)$, for some suitable r that is reasonably small but captures more than just one or two observations. The second variance is similar, except that now y is fixed so x is replaced by $\widehat{fp}$ and c is the smallest value exceeded by a proportion y of the sample from population P. The asymptotic normality in both cases enables a $100(1 - \alpha)\%$ confidence interval to be calculated as the appropriate estimate (i.e., $\widehat{tp}$ in the first case or $\widehat{fp}$ in the second) $\pm z_{1-\frac{\alpha}{2}}$ times the square root of the estimated variance [noting that $\alpha/2 = 1 - \Phi(z_{1-\frac{\alpha}{2}})$].

3.4.2 Parametric curve fitting

If a specific distribution has been assumed for each of the two populations and the model parameters have been estimated by maximum likelihood, then the asymptotic joint distribution of the parameters is obtained from standard maximum likelihood theory. Specifically, if $\boldsymbol{\theta} = (\theta_1, \ldots, \theta_p)^T$ is the vector of p unknown model parameters and $\log \mathcal{L}$ is the log-likelihood of the sample, then the maximum likelihood estimator $\hat{\boldsymbol{\theta}}$ is asymptotically distributed as a p-vector normal variate with mean $\boldsymbol{\theta}$ and dispersion matrix $\mathcal{I}^{-1}$ where the (i,j)th element of $\mathcal{I}$ is given by $-E\left(\frac{\partial^2 \log \mathcal{L}}{\partial \theta_i \partial \theta_j}\right)$. The parametric estimate of the ROC curve is a function of all the estimated model parameters, so we can use the above result, the fact that the two samples are independent, and then the delta method to obtain the variance and hence standard error of any point on the ROC curve. Unknown parameters are again replaced by their maximum likelihood estimates, and standard normal theory is then employed to obtain confidence intervals. However, it almost goes without saying that such an approach can involve very heavy mathematical manipulation and subsequent calculation.

While the many different distributional possibilities preclude a single specification of the $100(1-\alpha)\%$ confidence interval in the above case, the binormal model allows a relatively simple statement. For this model, the value y at the fp value x is given by $y = \Phi(\hat{a} + \hat{b}z_x)$ where $\hat{a}$ and $\hat{b}$ are the maximum likelihood estimates of parameters a and b, as obtained, for example, at the conclusion of the iterative process of Dorfman and Alf (1969). Now the variance of $\hat{a} + \hat{b}z_x$ is given by

$$V = \text{var}(\hat{a}) + z_x^2 \text{var}(\hat{b}) + 2z_x \text{cov}(\hat{a}, \hat{b}),$$

and this can be obtained easily because the variances of $\hat{a}, \hat{b}$ and the covariance between them are the elements of the inverse of the hessian matrix evaluated at $\hat{a}, \hat{b}$. $100(1-\alpha)\%$ confidence limits L and U for $a + bz_x$ are then given by $\hat{a} + \hat{b}z_x \mp V^{1/2} z_{1-\frac{\alpha}{2}}$, and this gives pointwise confidence limits at specified x values for $y = \Phi(a + bz_x)$ as $\Phi(L)$ and $\Phi(U)$. Ma and Hall (1993) extend pointwise confidence limits to confidence bands for the whole ROC curve.

Hsieh and Turnbull (1996) establish asymptotic normality, and derive explicit formulae for the covariance matrices, for all the estimators that they propose. In fact, the same asymptotic distribution obtains for Dorfman and Alf's categorized approach, for the regression approximation to it, and for the generalized least squares estimator. The minimum distance approach, however, has a different covariance matrix. Cai and Moskowitz (2004) likewise give covariance matrices for their estimators, which along with asymptotic joint normality leads to straightforward construction of confidence intervals.

3.4.3 Nonparametric estimation

Zou et al. (1997) derive confidence intervals for tp at a fixed value of fp, and also for (fp, tp) at a fixed value of the threshold t of the classification statistic, following their kernel-based nonparametric estimate of the ROC curve. In order to derive the intervals they first use logit transforms of all probabilities, on the grounds that a normal approximation works better on the transformed values. They thus use standard normal-based confidence intervals for the transformed values, and then use the inverse transform on the limits of the intervals to obtain confidence intervals for the original probabilities. The logit transform of a probability θ is

$$\phi = logit(\theta) = \log\left(\frac{\theta}{1-\theta}\right),$$

so that the inverse transform is

$$\theta = logit^{-1}(\phi) = \frac{1}{1+e^{-\phi}}.$$

In the present case we use $u = logit(fp), v = logit(tp), \hat{u} = logit(\widehat{fp})$, and $\hat{v} = logit(\widehat{tp})$.

For vertical confidence intervals on tp at fixed fp, $\hat{v}$ is asymptotically normally distributed with mean tp and estimated variance

$$s^2(\hat{v}) = \left\{\frac{1}{n_P}\widehat{tp}(1-\widehat{tp}) + \frac{1}{n_N}\left[\frac{\hat{g}(c)}{\hat{f}(c)}\right]^2 fp(1-fp)\right\}/[\widehat{tp}(1-\widehat{tp})]$$

where $\widehat{tp} = \hat{G}(c)$ and c is determined by $fp = \hat{F}(c)$. This gives the vertical $100(1-\alpha)\%$ confidence interval for v as $\hat{v} \pm s(\hat{v})z_{1-\frac{\alpha}{2}}$ so taking the inverse logit transform of the end points of this interval gives the vertical $100(1-\alpha)\%$ confidence interval for tp at fixed fp.

3.5. ESTIMATING SUMMARY INDICES

Turning to a confidence interval for (fp, tp) at a fixed value of the threshold t of the classification statistic, the pair of logit-transformed estimates $\hat{u}, \hat{v}$ are asymptotically independently normal with means u, v and variances estimated by $s^2(\hat{u}) = 1/[n_N \widehat{fp}(1 - \widehat{fp})]$ and $s^2(\hat{v}) = 1/[n_P \widehat{tp}(1 - \widehat{tp})]$ respectively. $100(1-\alpha)$ confidence intervals for each of u and v are thus given by $\hat{u} \pm s(\hat{u})z_{1-\frac{\alpha}{2}}$ and $\hat{v} \pm s(\hat{v})z_{1-\frac{\alpha}{2}}$, and since $\hat{u}, \hat{v}$ are independent then these intervals define a rectangular confidence region for (u, v) having confidence coefficient $100(1-\alpha)^2$. Applying the inverse logit transform to each corner of this rectangle gives a corresponding rectangular $100(1-\alpha)^2\%$ confidence region for (fp, tp) at the chosen value of t. Figure 3.6 shows 95% confidence rectangles calculated in this way for several threshold points of the nonparametric kernel-based ROC curve (Figure 3.5) for the quadratic classifier on the Pima Indian data.

Neither Peng and Zhou (2004) nor Ren *et al.* (2004) discuss the construction of confidence intervals for their methods, but rather they investigate their accuracies via mean squared error computations, both analytically and using Monte Carlo simulation.

3.4.4 Binary regression methods

Lloyd (2002) comments that no formulae are available for either mean or variance of his semi-parametric estimator, and Qin and Zhang (2003) likewise state that lack of analytical expressions precludes the derivation of formulae for confidence intervals. Both papers therefore recommend using bootstrap methods for computing such intervals, and both set out appropriate schemes for carrying out the bootstrap resampling.

3.5 Estimating summary indices

Estimating and plotting the ROC curve leads to graphical representations, but if we need a quantification of the information for use in further analysis we would typically turn to one of the available summary indices. Several such indices were defined for the theoretical ROC curve in Chapter 2, so we now outline their evaluation from sample data as well as discussing an extra data-based index.

Figure 3.6: 95% confidence rectangles for some threshold points of the nonparametric kernel-based ROC curve for the quadratic classifier on the Pima Indian data

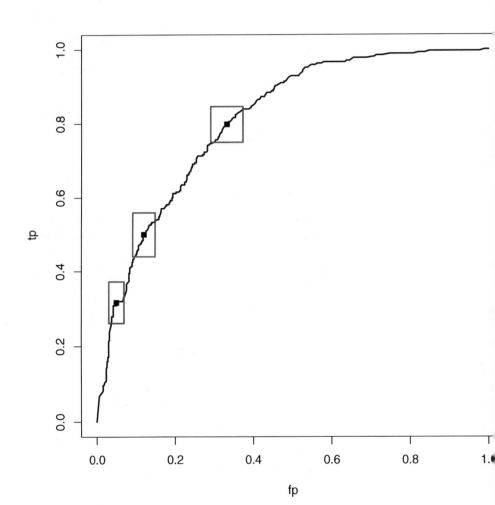

3.5.1 Area under the curve (AUC)

Estimation

The area under the curve has been defined in Section 2.4.1 as

$$\text{AUC} = \int_0^1 y(x)dx.$$

To estimate this quantity from sample data, we can estimate the ROC curve $y(x)$ either empirically or by fitting a smooth curve using one of the methods above, and then obtain $\widehat{\text{AUC}}$ by numerical integration. The latter can be effected by the trapezium rule—namely splitting the range $(0,1)$ into a number of subintervals, approximating the curve on each interval by the trapezium formed by its ends and the x-axis, and summing the areas of the trapezia—although this systematically produces a (small) underestimate of ROC curves (which are largely concave; Hanley and McNeil, 1982). An alternative suggested by Zou et al. (1997) is to use values of the curve at midpoints of subintervals times subinterval widths, which will tend to slightly overestimate the true area.

If the ROC curve has been estimated empirically, then numerical integration is unnecessary in view of the property established in Section 2.4.1 that AUC is equal to the probability that a randomly chosen individual from population P will yield a larger value of the classification statistic than a randomly chosen individual from population N. Translating this property from population probabilities to sample proportions leads to the equivalence first noted by Bamber (1975) that the area under the empirical ROC curve is equal to the Mann-Whitney U-statistic. A formal definition of this statistic is

$$U = \frac{1}{n_N n_P} \sum_{i=1}^{n_N} \sum_{j=1}^{n_P} [I(s_{Pj} > s_{Ni}) + \frac{1}{2} I(s_{Pj} = s_{Ni})],$$

where $I(\mathcal{S}) = 1$ if $\mathcal{S}$ is true and 0 otherwise. In words, if we consider all possible pairs of individuals, one from each sample, then U is the sum of the proportion of pairs for which the score for the individual from sample P exceeds that for the individual from sample N, and half the proportion of ties. Since the classification statistic is continuous, then providing that the scale of measurement is sufficiently accurate we can assume a negligible probability of obtaining ties, in which case it follows that $E(U) = p(S_P > S_N) = \text{AUC}$ so that U provides an unbiased estimator of AUC. It is worth noting (although outside the remit of this book) that if the classification statistic is only measured on an ordinal scale, sometimes referred to as a *rating* scale, then there is a non-zero probability of observing ties and U is a biased estimator of AUC. It is also worth noting that various other standard two-sample rank statistics can be written as functions of the empirical ROC curve (see Pepe, 2003, page 127); however, since the Mann-Whitney U directly equals AUC it is the one generally adopted.

In the case of parametric models based on specific distributional assumptions, AUC can usually be derived analytically in terms of the population parameters so is estimated simply by substituting either sample analogues or maximum likelihood estimates for these parameters. In particular, for the binormal model (and hence also for the parametric assumption of normality of each separate population) we saw in Section 2.5 that

$$\text{AUC} = \Phi\left(\frac{\mu_P - \mu_N}{\sqrt{\sigma_P^2 + \sigma_N^2}}\right) = \Phi\left(\frac{a}{\sqrt{1+b^2}}\right),$$

where $a = (\mu_P - \mu_N)/\sigma_P$ and $b = \sigma_N/\sigma_P$. If populations N and P are assumed to be normal from the outset, then AUC can be estimated by substituting means and standard deviations of the two samples into the first expression, although a preliminary Box-Cox transformation is advisable to ensure consonance of samples with normality assumptions. If the weaker assumptions of the binormal model are made, then AUC is estimated by substituting estimates of a and b from one of the previously described estimation methods into the second expression. If a nonparametric method, such as the one by Zou *et al.* (1997) has been used to estimate the ROC curve then numerical integration is generally needed to estimate AUC. However, if a normal kernel is used instead of the biweight kernel of Section 3.3.3, then Lloyd (1998) has shown that the resulting kernel estimate of AUC can be expressed as

$$\widehat{\text{AUC}} = \frac{1}{n_N n_P} \sum_{i=1}^{n_N} \sum_{j=1}^{n_P} \Phi\left(\frac{s_{Pj} - s_{Ni}}{\sqrt{h_N^2 + h_P^2}}\right).$$

Faraggi and Reiser (2002) report the results of a simulation study in which they generated data under a large number of different population structures and different AUC conditions, and then compared all the above methods of estimating AUC with respect to bias and mean squared error of estimates. This followed a previous, more limited simulation study by Hajian-Tilaki *et al.* (1997) comparing the performance of the Mann-Whitney U with the maximum-likelihood binormal method of Metz *et al.* (1998). Broadly, the findings suggest that there is not much to choose between the methods for symmetric and normal-like populations, that transformation to normality plus parametric normal assumptions performs well for skewed populations, but that Mann-Whitney U and nonparametric methods strongly outperform the others once the populations exhibit bimodality or mixtures of distributions. Even though the Mann-Whitney U was not usually best in terms of mean squared error it was often close to best, but was overtaken by kernel methods for poorly separated populations that contained mixtures.

Confidence intervals

In order to obtain confidence intervals for the true AUC, we need to have an estimate of the variance of $\widehat{\text{AUC}}$. For parametric and semiparametric methods, maximum likelihood theory will yield asymptotic expressions for the variances and covariances of the parameters and so the delta method will yield the required variance (as with confidence intervals for points on the ROC curve above). We therefore focus here on the nonparametric approaches.

Consider first the empirical ROC curve and the Mann-Whitney U as estimate of AUC. Given the standard nature of the statistic, and hence its central role in ROC analysis, various authors have derived and discussed asymptotic expressions for its variance including Bamber (1975), Hanley and McNeil (1982), DeLong *et al.* (1988), Hanley and Hajian-Tilaki (1997), and Pepe (2003). Perhaps the most familiar expression is the one given by Hanley and McNeil (1982):

$$s^2(\widehat{\text{AUC}}) = \frac{1}{n_N n_P}(\text{AUC}[1 - \text{AUC}] + [n_P - 1][Q_1 - \text{AUC}^2] + [n_N - 1][Q_2 - \text{AUC}^2]),$$

where Q_1 is the probability that the classification scores of two randomly chosen individuals from population P exceed the score of a randomly chosen individual from population N, and Q_2 is the converse probability that the classification score of a randomly chosen individual from population P exceeds both scores of two randomly chosen individuals from population N. Sample estimates of Q_1 and Q_2 are the obvious ones; for example $\widehat{Q}_1$ is given by considering all possible triples comprising two sample members from P and one from N, and finding the proportion for which the two P scores exceed the N score. Substituting $\widehat{\text{AUC}}, \widehat{Q}_1,$ and $\widehat{Q}_2$ in the right-hand side above gives an estimate of $s^2(\widehat{\text{AUC}})$, whence a $100(1-\alpha)\%$ confidence interval for AUC is given by $\widehat{\text{AUC}} \pm s(\widehat{\text{AUC}}) z_{1-\frac{\alpha}{2}}$.

The alternative approach due to DeLong et al. (1988) and exemplified by Pepe (2003) gives perhaps a simpler estimate, and one that introduces the extra useful concept of a placement value. The placement value of a score s with reference to a specified population is that population's survivor function at s. Thus the placement value for s in population N is $1 - F(s)$, and for s in population P it is $1 - G(s)$. Empirical estimates of placement values are given by the obvious proportions. Thus the placement value of observation s_{Ni} in population P, denoted s_{Ni}^P, is the proportion of sample values from P that exceed s_{Ni}, and $\text{var}(s_{Ni}^P)$ is the variance of the placement values of each observation from N with respect to population P. Likewise, $\text{var}(s_{Pi}^N)$ is the variance of the placement values of each observation from P with respect to population N. The DeLong et al. (1988) estimate of variance of $\widehat{\text{AUC}}$ is given in terms of these variances:

$$s^2(\widehat{\text{AUC}}) = \frac{1}{n_P}\text{var}(s_{Pi}^N) + \frac{1}{n_N}\text{var}(s_{Ni}^P).$$

With regard to kernel estimates of AUC, Zou et al. (1997) comment that the smoothing introduced by the kernel estimate does not affect the true standard error to a first order of approximation, so the above values of $s^2(\widehat{\text{AUC}})$ may be used also for the area under the smoothed curve. They suggest, however, that any confidence intervals should be computed for transformed $\widehat{\text{AUC}}$ and then back-transformed to the original scale, in order to ensure that they stay within the $(0,1)$ range. They use the transformation $-\log(1-\text{AUC})$, and derive a $100(1-\alpha)\%$ confidence interval for AUC as $1 - (1 - \widehat{\text{AUC}})\exp[\pm z_{1-\frac{\alpha}{2}} s(\widehat{\text{AUC}})/(1-\widehat{\text{AUC}})]$.

In a recent contribution, Qin and Zhou (2006) propose an empirical likelihood approach for inference about AUC. They define an empirical likelihood ratio for AUC, show that its limiting distribution is scaled chi-squared, and then obtain a confidence interval for AUC using this distribution. It is shown by simulation that this method generally outperforms several other methods, particularly as AUC approaches 1.

3.5. ESTIMATING SUMMARY INDICES

Finally, Cortes and Mohri (2005) use combinatorial approaches to derive expressions for the expectation and variance of AUC that require only the input values of simple parameters such as error rate and sample sizes from each of the two populations. They then show that if a confidence interval is available for the error rate of the classifier over a sample, a corresponding confidence interval can be derived for AUC with the help of the above expressions.

3.5.2 Partial area under the curve (PAUC)

From Section 2.4.2, the partial area under the curve between values f_1 and f_2 of fp is given by

$$\text{PAUC}(f_1, f_2) = \int_{f_1}^{f_2} y(x)dx.$$

Parametric estimation of this partial area is achieved by assuming specific distributional forms for the classification scores in each population, estimating the parameters of these distributions and hence estimating the ROC curve as in Section 3.3.2 above, and then integrating this curve numerically between f_1 and f_2 to obtain the required estimate. We have already seen that under the assumption of a binormal model, or alternatively of separate normal distributions for the two populations, the ROC curve is given by $y(x) = \Phi(a + bz_x)$ so that

$$\widehat{\text{PAUC}}(f_1, f_2) = \int_{f_1}^{f_2} \Phi(\hat{a} + \hat{b}z_x)dx,$$

where $\hat{a}, \hat{b}$ are estimates of a, b as described in Section 3.3.2. This estimator was first discussed by McClish (1989).

More recently, interest has turned to nonparametric estimation, and Dodd and Pepe (2003b) and Liu *et al.* (2005) have independently proposed essentially the same estimator. Both derivations start by showing that the partial area under the curve is equal to the joint probability that a randomly chosen individual from population P is greater than a randomly chosen individual from population N *and the latter individual falls within the relevant population quantiles.* Thus, whereas AUC = $p(S_P > S_N)$, PAUC$(f_1, f_2) = p(S_P > S_N, f_1 \leq 1 - F(S_N) \leq f_2)$. A sample estimate of this quantity is fairly straightforward, namely the proportion of all possible pairs of observations, one from each population, that satisfy the required constraint and for which the observation from P exceeds the one from N. Pairs (s_{Ni}, s_{Pj}) satisfy the constraint if the rank r_j of s_{Pj} among the sample from P, i.e., $r_j = \sum_{k=1}^{n_P} I(s_{Pk} \leq s_{Pj})$, is such that $f_1 \leq \frac{r_j}{n_P} \leq f_2$. Thus the estimator can be formally written

$$\widehat{\text{PAUC}}(f_1, f_2) = \frac{1}{n_N n_P} \sum_{i=1}^{n_N} \sum_{j=1}^{n_P} I(s_{Pj} > s_{Ni}) I(f_1 \leq \frac{r_j}{n_P} \leq f_2)$$

where $I(.)$ is defined alongside the Mann-Whitney U in Section 3.5.1 above. We also see directly that if $f_1 = 0, f_2 = 1$ and there are no ties (in the continuous data), then this estimator reduces to the Mann-Whitney U.

Liu et al. (2005) derive the asymptotic distribution of the estimator, give consistent estimates of its variance, and hence obtain confidence intervals for PAUC(f_1, f_2). Dodd and Pepe (2003b) recommend the use of bootstrap methods for obtaining confidence intervals. They also investigate by simulation the small sample properties and robustness of the estimator as well as the coverage probabilities of the computed intervals. They conclude that the nonparametric estimator is substantially more robust than the parametric normal-based estimator, and so better for general-purpose use. The latter is the most efficient estimator in the presence of normality of data but produces biased estimates, and the relatively small loss in efficiency of the nonparametric method is more than made up for by its greater robustness to nonnormality. Wieand et al. (1989) give a nonparametric estimate and its variance for the generalized summary measure outlined in Section 2.4.2 that includes AUC and PAUC as special cases.

3.5.3 Optimal classification threshold

In Section 2.3.1 we saw that if $c(\text{P}|\text{N})$ and $c(\text{N}|\text{P})$ are the costs incurred when an N individual is allocated to P and when a P individual is allocated to N, respectively, and if the relative proportions of P and N individuals are q and $1 - q$, then the threshold that minimizes the expected cost due to misallocation (the quantity C of Section 2.3.1) is the threshold that corresponds to slope

$$\frac{dy}{dx} = \frac{(1-q)c(\text{P}|\text{N})}{qc(\text{N}|\text{P})}$$

of the (theoretical) ROC curve. Estimation of this optimal threshold with sample data is relatively straightforward if we have obtained a smooth estimate of the ROC curve: it is the point at which the line with the above slope is a tangent to the curve. If the ROC curve has been estimated empirically, however, it will be a step function. In this case Zweig and Campbell (1993) define the estimate as the point at which a line with the above slope first intersects the empirical ROC curve when the line is moved from the upper left-hand corner of the plot towards the curve.

3.5.4 Loss difference plots and the LC index

Adams and Hand (1999) have further studied the expected cost due to misclassification of an individual, C, but they term it the expected *loss* of the classifier. They first quote the standard result in classification that minimum loss is achieved at threshold

$$t = \frac{c(\text{P}|\text{N})}{[c(\text{N}|\text{P}) + c(\text{P}|\text{N})]} = \frac{1}{1+k},$$

where $k = c(\text{P}|\text{N})/c(\text{N}|\text{P})$ is the cost ratio. Thus the important quantity in assessment of classifiers is the ratio k rather than the separate costs. The loss associated with any point on the ROC curve is the length of projection of that point onto the line with slope $-\frac{(1-q)c(\text{P}|\text{N})}{qc(\text{N}|\text{P})}$ through $(0,1)$ as origin. For a given k, the threshold minimizing the loss corresponds to that position on the ROC curve for which this projection is smallest. Thus if k is known, then an awkward projection operation is required to determine the overall loss. On the other hand, if k is not known then AUC is often used as a measure of performance of a classifier. However, if two classifiers are being compared and their ROC curves cross, then one classifier will be superior for some values of k but inferior for other values. If the analyst has some idea as to which side of the crossing point the cost ratio is likely to fall, then comparing classifiers by integrating over all cost ratios to find AUC will be misleading and indeed pointless.

This consideration led Adams and Hand (1999) to propose an alternative graphical representation to ROC and an alternative summary measure to AUC, that would take advantage of any knowledge available about the cost ratio. They termed the new graph a *loss difference* plot. It consists simply of a plot of the sign of the differences in loss for the two classifiers against $1/(1+k)$ (so that the scale of the horizontal axis is $(0,1)$ rather than $(0,\infty)$ if k is used). Furthermore, since only their ratio is relevant the individual costs can be arbitrarily rescaled, and rescaling so that $c(N|P) + c(P|N) = 1$ implies that $\frac{1}{1+k} = c(P|N) = c$, say. Thus values $c \in (0, 0.5)$ mean that misclassifying an individual from P is less serious than misclassifying one from N, while values $c \in (0.5, 1)$ mean the converse. The loss difference plot gives a partition of $(0, 1)$ into segments showing how the classifiers alternate in terms of their superiority.

To derive a new summary index, Adams and Hand (1999) first required a quantification of the degree of confidence that could be ascribed to a particular value of c (as would typically be done by an expert in the substantive matter of the application). If the expert assesses the most feasible interval for k to be (a, b), with its most likely value to be at $k = m$, then Adams and Hand suggest using a triangular distribution $D(c)$ for the degree of confidence, with end points at $c = 1/(1 + a)$ and $1/(1 + b)$ and apex at $c = 1/(1 + m)$. To obtain their new index for comparing two classifiers, which they call the LC (Loss Comparison) index, they define a function $L(c)$ which takes value $+1$ for those regions of c where A is superior (its loss plot lying below that of B) and value -1 for those regions where B is superior. The LC index of superiority of A over B is then the integral of the product $L(c)D(c)$ over $(0, 1)$, and this gives a measure of confidence in the range $(-1, 1)$ that A will yield a smaller loss than B.

To illustrate the loss difference plot and the LC index, and to compare them with the ROC curve and AUC, return to the Pima Indian data previously considered in Section 3.3.1. Figure 3.2 showed the ROC curve for the quadratic discriminant function and for a neural network classifier. The AUCs for these classifiers are respectively 0.7781 and 0.7793, suggesting that there is little to choose between them but that if anything the neural network is superior. However, the two ROC curves cross; so superiority of one or other classifier will depend on relative costs, and such information is not easily incorporated into the diagram. For example, suppose that one is confident that k lies in $(0.1, 0.25)$ with a most likely value of 0.14 (i.e., misclassifying a diabetic as nondiabetic is between 4 and 10 times as serious as the converse, and most likely to be 7 times as serious). There is no immediate way in which we could incorporate this information into Figure 3.2. However, Figure 3.7 shows the loss difference plot with superimposed distribution $D(c)$. Values $0.1, 0.14, 0.25$ of k translate to $0.80, 0.875, 0.91$ of c, which determine the triangle for $D(c)$, while the bold line segments above the horizontal axis show the regions of c for which each classifier is superior. The LC

3.5. ESTIMATING SUMMARY INDICES

Figure 3.7: Loss difference plot with superimposed triangular distribution for the Pima Indian data

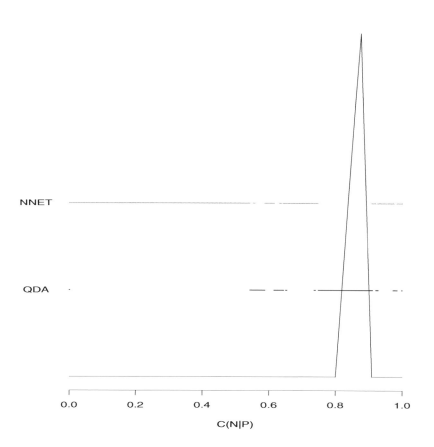

index is -1.0, which very strongly suggests that the quadratic classifier is superior. While the neural network is superior for the majority of values of c, the quadratic classifier is superior in the crucial region where the expert assesses the cost ratio to lie. Hence a very different result is obtained with LC than with AUC.

The LC plot and other variants of ROC curves are discussed further in Section 8.2.

3.6 Further reading

The present book is concerned exclusively with continuous classification scores, so little is mentioned about the other possible types such as binary, nominal, or ordinal scores. While the binary and nominal cases give rise to their own specific problems and techniques, much of the theory for continuous scores is echoed in the ordinal case, and reading about the latter is helpful for understanding many of the continuous results. Both Pepe (2003) and Zhou *et al.* (2002) contain extensive accounts, and are recommended for readers who wish to establish the various connections.

Another topic alluded to above, but outside the scope of the present text, is the elicitation of costs or cost ratios for use in optimal threshold determination, loss difference plots, and the LC index. This topic is of critical concern in medicine, so has attracted study in this area; see Gold *et al.* (1977) and Weinstein *et al.* (1996), for example.

Placement values have been used in order to obtain relatively simple expressions for variances of empirical estimates of AUC, but they are useful more widely; see Pepe (2003) for details. The concept has also been extended to *restricted* placement values, which have connections with PAUC analogous to those between placement values and AUC (Dodd and Pepe, 2003b).

Chapter 4

Further inference on single curves

4.1 Introduction

In the previous chapter we considered some inferential aspects associated with single ROC curves, but there we focussed exclusively on estimation. It was assumed that empirically gathered data sets constitute samples from two populations (P and N), that for a given classifier there exists a "true" ROC curve representing the scores of individuals from these populations, and that the sample data will be used to estimate either this ROC curve itself or some summary characterization of it such as AUC or PAUC. A variety of assumptions about the population scores can be made, and hence a variety of different estimators and estimation methods were reviewed. As well as finding estimates of the relevant parameters it is important to quantify the variability of these estimates from sample to sample, so this led to consideration of confidence intervals for the various parameters.

In this chapter, we consider some remaining inferential matters for single ROC curves. The main gap to be filled concerns the testing of hypotheses, but other important topics include considerations of the sample sizes necessary to ensure that clear conclusions can be reached, and the effects that errors in measurement can have on the inferential procedures. While it is probably true that most interest in statistical analysis resides in estimation, and indeed that simple hypotheses against two-sided alternatives can usually be tested by checking whether or not the hypothesized values lie inside appropriately constructed confidence intervals, nevertheless there are many situations in which specific hypothesis testing procedures are necessary. This is equally true for ROC curve analysis, and in this chapter we review such hypothesis tests; situations for which no specific tests have been given can generally be tackled using the confidence interval approach, with appropriate reference to Chapter 3. The asymptotic theory that provides a foundation for some of the mathematical derivations is covered by Hsieh and Turnbull (1996), Girling (2000), and references cited therein.

4.2 Tests of separation of P and N population scores

The ROC curve provides a description of the separation between the distributions of the classification score S in the two populations. We know from Section 2.2.2 that the chance diagonal, namely the line joining the points $(0,0)$ and $(1,1)$, represents the uninformative ROC curve for which the probability of allocating an individual to population P is the same whether that individual has come from population P or population N. So to measure the extent of separation between the population scores, we need to measure (in some way that can be calculated from sample data) the difference between the ROC curve and the chance diagonal. A test of separation of the two populations of scores can then be built from this measure of difference for the samples. Various measures can be defined, and we consider below the two most common ones for practical applications. The first is a direct measure of difference between the ROC curve and the chance diagonal, and the second is the difference between the areas under the curve and the line.

4.2.1 Kolmogorov-Smirnov test for the empirical ROC

It was noted in Section 2.4.3 that a very simple measure of difference between the ROC curve and the chance diagonal is the maximum vertical distance MVD between them across their range $(0, 1)$. MVD is thus an indicator of how far the curve deviates from "randomness," and ranges from 0 for an uninformative curve to 1 for a perfect discriminator. Now we know from Section 2.2.4 that the ROC curve has equation
$$y = 1 - G[F^{-1}(1-x)] \quad (0 \leq x \leq 1),$$

4.2 TESTS OF SEPARATION OF POPULATION SCORES

where F is the distribution function of S in population N and G is the distribution function of S in population P, while the chance diagonal has equation $y = x$ for $0 \leq x \leq 1$. Thus we can write

$$\text{MVD} = \max_x |1 - G[F^{-1}(1-x)] - x| = \max_x |(1-x) - G[F^{-1}(1-x)]|.$$

But if $t = F^{-1}(1-x)$ then $(1-x) = F(t)$, and since $F(\cdot)$ is a distribution function then t ranges over $(-\infty, \infty)$. Thus by direct substitution into the last expression above, we see that

$$\text{MVD} = \max_t |F(t) - G(t)| = \sup_{t \in (-\infty, \infty)} |F(t) - G(t)|.$$

If we now have sample data from which we have obtained the empirical ROC curve

$$y = 1 - \hat{G}[\hat{F}^{-1}(1-x)] \quad (0 \leq x \leq 1)$$

as described in Section 3.3.1, then the maximum vertical distance $\widehat{\text{MVD}}$ between this curve and the chance diagonal is

$$\widehat{\text{MVD}} = \sup_{t \in (-\infty, \infty)} |\hat{F}(t) - \hat{G}(t)|.$$

Thus $\widehat{\text{MVD}}$ is just the well-known Kolmogorov-Smirnov statistic for testing nonparametrically the equality of the two population distribution functions F and G. The statistic is so named because Kolmogorov first devised a test for consonance of an empirical distribution function with a population distribution function, while Smirnov extended it to a test of difference between two empirical distribution functions. Most textbooks on nonparametric methods (e.g., Sprent, 1989) give full details of the test as well as tables of critical values, which can be used directly for testing the significance of $\widehat{\text{MVD}}$ in ROC analysis. Equivalence of MVD and the Youden Index, YI, established in Section 2.4.3, means that the same tables can be used for testing the significance of $\widehat{YI}$.

4.2.2 Test of AUC = 0.5

The maximum vertical distance statistic MVD measures the distance between the ROC curve and the chance diagonal at their point of greatest separation, but pays no attention to their difference at other points in (0, 1). An admittedly degenerate albeit plausible situation would be one in which the ROC curve lay along the chance diagonal for most of this range, but had one "jump" with a substantial MVD followed by a small region before it returned to the chance diagonal. Such a situation would yield a potentially misleading significant result, which would not have arisen had the measure of separation used information from all points in (0, 1). One way of utilizing all this information is via AUC, the area under the curve. A simple way of testing whether the classifier under study has any ability to discriminate between populations P and N is therefore to test whether its AUC is significantly greater than the area under the chance diagonal, i.e., 0.5. The one-sided hypothesis seems appropriate, as the classifier will only by useful if its ROC lies uniformly *above* the chance diagonal.

The null and alternative hypotheses are therefore H_0 : AUC = 0.5 and H_a : AUC > 0.5, and a natural test statistic would seem to be

$$Z = \frac{\widehat{\text{AUC}} - 0.5}{s(\widehat{\text{AUC}})}$$

where $s^2(\widehat{\text{AUC}})$ is the estimated variance of the estimated AUC. The various possible approaches to estimating AUC, and their consequent variances, were outlined in Section 3.5.1. On employing asymptotic arguments (as in Hseih and Turnbull, 1996, or Girling, 2000) the distribution of Z under the null hypothesis can be approximated by a standard normal distribution, so the usual procedure for carrying out a test under normality can be followed.

4.3 Sample size calculations

While most derivations of inferential procedures assume fixed sample sizes and focus on the production of a formula for either a confidence interval or a test statistic, in actual practice such formulae are often used in an inverse fashion at the planning stage to determine the sizes of samples necessary either to achieve a given precision in confidence interval estimation or to detect a given departure from null conditions in hypothesis testing. In ROC curve analysis, such planning considerations generally focus on AUC: how large must the samples be to obtain an estimate of AUC that is sufficiently precise, or to test a hypothesis about AUC with at least a specified power?

4.3.1 Confidence intervals

We saw in Section 3.5.1 that if $\widehat{\text{AUC}}$ is an estimate of AUC, then a $100(1-\alpha)\%$ confidence interval for AUC is given by $\widehat{\text{AUC}} \pm s(\widehat{\text{AUC}})z_{1-\frac{\alpha}{2}}$ where $s^2(\widehat{\text{AUC}})$ is the estimated variance of $\widehat{\text{AUC}}$. If AUC is estimated parametrically, then formulae for $s^2(\widehat{\text{AUC}})$ are usually derivable from the specific parametric model being used so we do not provide a general method for sample size determination in these cases. However, if the estimation is done either empirically or nonparametrically, then Hanley and McNeil (1982) (following Bamber, 1975) have derived the general formula

$$s^2(\widehat{\text{AUC}}) = \frac{1}{n_N n_P}(\text{AUC}[1-\text{AUC}] + [n_P - 1][Q_1 - \text{AUC}^2] + [n_N - 1][Q_2 - \text{AUC}^2]),$$

where Q_1 is the probability that the classification scores of two randomly chosen individuals from population P exceed the score of a randomly chosen individual from population N, and Q_2 is the converse probability that the classification score of a randomly chosen individual from population P exceeds both scores of two randomly chosen individuals from population N.

Thus the question of interest boils down to asking: "What are the smallest values of n_N and n_P that will ensure that $s^2(\widehat{\text{AUC}})$ lies below a stated tolerance level?" If we knew the values of AUC, Q_1, and Q_2, then we would obtain suitable values of n_N and n_P by simple trial and error. However, remember that this exercise is being conducted at planning stage, so not only do we not know any of these values, we do not even (yet) have the data from which they can be estimated! Consequently, a certain amount of guesswork and approximation is needed. We probably have a rough idea as to the sort of classifier we will use and hence of a likely target value of AUC, so we can use this in the formula. But what about Q_1 and Q_2, which are rather complex functions of the underlying populations being studied and will therefore differ between applications? Fortunately, Hanley and McNeil (1982) show that not only is the expression for $s^2(\widehat{\text{AUC}})$ almost entirely determined by AUC, and only very slightly affected by other parameters, but also that the relationship between the two is almost unchanged across a range of different population models. Using the simplest and most conservative model, therefore, Hanley and McNeil (1982) suggest approximating Q_1 by AUC/(2 - AUC) and Q_2 by $2\text{AUC}^2/(1 + \text{AUC})$. Substituting all these values into the formula above reduces it to a simple function of n_N and n_P, from which suitable sample sizes that meet the required tolerance level can be deduced by trial and error.

4.3.2 Hypothesis tests

Suppose now that we wish to set up an experiment to determine whether the (true) ROC curve of a classifier under study has a satisfactory value of AUC or not, and we need to determine the minimum sample sizes in the two populations in order to obtain conclusive results. Two typical scenarios might arise if we are developing a new classifier for some particular application and want to establish whether either (i) the classifier is worthwhile (i.e., has a better performance than random allocation), or (ii) the classifier exceeds some specified quality level of performance. Both scenarios can be dealt with under a hypothesis-testing framework, in which we test the null hypothesis that the classifier's AUC equals some specified value θ_0, say, against the alternative that it is greater than θ_0. In the former case we would set $\theta_0 = 0.5$, and if the null hypothesis is rejected then we have evidence that the classifier is better than random allocation, while in the latter case we would set θ_0 equal to a value that reflects the minimum acceptable level of performance, say 0.85, and if the null hypothesis is rejected then we have evidence that the classifier's performance exceeds this level.

Thus, in general, the null and alternative hypotheses are H_0 : AUC $= \theta_0$ and H_a : AUC $> \theta_0$, and following the reasoning in Section 4.2.2 we would base the test on the statistic

$$Z = \frac{\widehat{\text{AUC}} - \theta_0}{s(\widehat{\text{AUC}}|\theta_0)}$$

where $s^2(\widehat{\text{AUC}}|\theta_0)$ is the estimated variance of the estimated AUC when the true AUC equals θ_0. Since we are testing the null hypothesis against a one-sided alternative, we reject the null hypothesis if the calculated value of Z exceeds some constant that is appropriate for the desired significance level of the test. Using the approximate normality of Z, therefore, for a significance level α we reject H_0 if the calculated value of Z exceeds $z_{1-\alpha}$ where, as before, $1 - \alpha = \Phi(z_{1-\alpha}) = \Pr(Z \leq z_{1-\alpha})$ so that $\Pr(Z > z_{1-\alpha}) = \alpha$. Thus we reject H_0 if

$$\frac{\widehat{\text{AUC}} - \theta_0}{s(\widehat{\text{AUC}}|\theta_0)} > z_{1-\alpha},$$

4.3. SAMPLE SIZE CALCULATIONS

in other words if

$$\widehat{\mathrm{AUC}} > \theta_0 + z_{1-\alpha} s(\widehat{\mathrm{AUC}}|\theta_0).$$

If the null hypothesis is true then $\widehat{\mathrm{AUC}}$ has (approximately) a normal distribution with mean θ_0 and variance $s^2(\widehat{\mathrm{AUC}}|\theta_0)$, so it follows directly that the significance level (i.e., the probability of rejecting the null hypothesis when it is true) is α.

At the outset of this section, our stated aim was to determine the minimum sample sizes required in the two populations in order to obtain conclusive results, so we must make more precise what we mean by "conclusive results." Ideally, we would like to be able to reject H_0 whenever AUC departs from θ_0, but we know this to be a practical impossibility. Of course, the more that AUC departs from θ_0, the more *likely* we are to reject H_0, but because of the vagaries of random sampling there is always a chance of making a mistake no matter how far AUC is from θ_0. So the best we can hope for is to ensure that we have a high probability—at least β, say—that we will reject H_0 whenever AUC equals or exceeds some value θ_1 that represents the minimum departure that we would like to detect. Since the probability of rejection will increase as AUC rises above θ_1, we can impose the precise requirement that the probability of rejecting H_0 when AUC $= \theta_1$, i.e., the *power* of the test at θ_1, be the specified value β.

But if AUC $= \theta_1$ then $\widehat{\mathrm{AUC}}$ has (approximately) a normal distribution with mean θ_1 and variance $s^2(\widehat{\mathrm{AUC}}|\theta_1)$, so the probability of rejecting H_0 is the probability that

$$\frac{\widehat{\mathrm{AUC}} - \theta_1}{s(\widehat{\mathrm{AUC}}|\theta_1)} > \frac{\theta_0 - \theta_1 + z_{1-\alpha} s(\widehat{\mathrm{AUC}}|\theta_0)}{s(\widehat{\mathrm{AUC}}|\theta_1)}$$

where the quantity on the left-hand side of the inequality has a standard normal distribution. However, whenever the desired power is greater than 0.5 (as it always will be), then the right-hand side of the inequality will necessarily be negative. Moreover, $\Pr(Z > -a) = \Pr(Z < a)$ for all a and standard normal Z, so for power β we require

$$-\left(\frac{\theta_0 - \theta_1 + z_{1-\alpha} s(\widehat{\mathrm{AUC}}|\theta_0)}{s(\widehat{\mathrm{AUC}}|\theta_1)}\right) = z_\beta$$

which can be re-expressed as

$$z_{1-\alpha} s(\widehat{\mathrm{AUC}}|\theta_0) + z_\beta s(\widehat{\mathrm{AUC}}|\theta_1) = (\theta_1 - \theta_0).$$

To determine sample sizes for the experiment, the researcher must first specify target values for the significance level α, the power β, and the desired magnitude $\Delta = \theta_1 - \theta_0$ of departure from the AUC null hypothesis that is to be detected. Using the relationship between sample sizes and $s^2(\widehat{\text{AUC}}|\theta)$, the variance at AUC $= \theta$, values of the former that provide the best approximation to the above equation can then be deduced. A direct way of doing this is by using the Hanley and McNeil (1982) formula in a numerical trial and error process, as given for the method with confidence intervals in Section 4.3.1, but other authors have proposed various approximations to obviate the necessity for trial and error. To simplify matters, target sample sizes n_N, n_P for the two populations are either taken as equal (so that $n_P = n_N$) or in a given ratio κ (so that $n_P = \kappa n_N$).

Pepe (2003) assumes a common variance at all values of θ and points out that if one can further assume a parametric form for the ROC curve, such as the binormal form, then an explicit expression can be obtained for the asymptotic variance and sample sizes can be deduced directly. Indeed, if specific distributions can be assumed for the populations N and P then large samples can be generated from these populations and the variance can be estimated using the placement values as discussed in the confidence interval subsection 3.5.1. Obuchowski and McClish (1997) consider binormal ROC curves, and derive expressions that can be used in sample size calculations for such situations, while Obuchowski (1998) gives expressions for the general case based on approximations to $s^2(\widehat{\text{AUC}}|\theta)$ in the form of a ratio (variance function)/(sample size). A summary of these results is given in Zhou *et al.* (2002). All these references also contain details about quantities other than AUC which may be the focus of sample size calculations, such as *fp*, *tp*, and PAUC.

4.4 Errors in measurements

ROC curve analysis features very prominently in medical and epidemiological studies, where the classifier is used to detect the presence of a disease. Frequently, such a classifier is a single variable whose value is determined either by machine reading or by chemical analysis in a laboratory, and it is known as a *diagnostic marker for* or a *diagnostic indicator of* the disease. Relating this scenario to our standard notation, population P is therefore the population of individuals *with* the disease

4.4. ERRORS IN MEASUREMENTS

("positive" marker or indicator) and population N is the population of individuals *without* it ("negative" marker or indicator), while the classifier is just the marker variable. Coffin and Sukhatme (1997) point out that the susceptibility of diagnostic markers to measurement error has been extensively documented in the medical literature. They give as published examples the use of systolic and diastolic blood pressure as markers for hypertension, and measurements on cerebrospinal fluid as markers for bacterial meningitis. These markers are subject to measurement errors that can be attributed to the laboratory equipment, to the technician who uses it, or to both causes. Of course, measurement errors can occur in many, perhaps most, application domains. They can arise in any area where there is propensity for changes in population values (e.g., financial studies such as when a bank wants to distinguish potential customers between those who will repay loans on time and those who will incur bad debts, and circumstances change over the period of the loan), or there is variation in laboratory conditions (e.g., reliability studies where electronic components to be assessed as either reliable or unreliable are tested in a variety of centers). However, it is undoubtedly in the medical context that the study of methods for handling errors in measurements has been prominent. In some applications the presence of measurement errors does not affect the method of analysis, particularly if the same error mechanism operates in both populations. So the derivation of a classifier, the calculation of the classification scores, and descriptive analysis of the system can proceed in the usual fashion. However, measurement errors will upset some of the properties of standard estimators of population quantities like AUC, and it is with this aspect that we are concerned here.

The general structure of the problem can be simply stated. We are interested in the AUC for a ROC curve obtained from classifier scores S_P, S_N in populations P and N, and estimate it by $\widehat{\mathrm{AUC}}$ as calculated from the empirical ROC curve based on values S_{Pi} ($i = 1, \ldots, n_P$) and S_{Ni} ($i = 1, \ldots, n_N$) observed in samples from the two populations. We have seen in Section 3.5.1 that $\widehat{\mathrm{AUC}}$ is an unbiased estimator of $\Pr(S_P > S_N)$ and hence of AUC. However, now suppose that the measurements have been made with error, so that instead of observing S_{Pi}, S_{Ni} we actually observe $S'_{Pi} = S_{Pi} + \epsilon_i$ ($i = 1, \ldots, n_P$) and $S'_{Ni} = S_{Ni} + \eta_i$ ($i = 1, \ldots, n_N$) where ϵ_i and η_i represent error terms. It can be readily seen that $\widehat{\mathrm{AUC}}'$ as calculated from these values is an unbiased estimator of $\Pr(S'_P > S'_N) = \mathrm{AUC}'$, say, but it will no longer be unbiased for AUC. So the main questions concern the nature and size of the bias, the effects this bias has on confidence intervals or tests for AUC, and corrections that should be applied to overcome these effects.

Coffin and Sukhatme (1995, 1997) were the first authors to recognize and tackle these problems. In the first paper they looked at the parametric case and in the second one the nonparametric case, focussing on finding expressions for and estimates of the bias. We here briefly summarize their approach to the nonparametric case. They assume that all ϵ_i and η_i are independent and identically distributed, coming from a population that has mean 0 and variance σ^2, and that, following our earlier notation, f, F are the density and distribution functions of population N while g, G are the corresponding functions for population P. Using a Taylor series expansion up to order 3, they show that

$$E(\widehat{AUC}') \approx AUC - \frac{1}{2}E(\delta^2)E[G''(S_N)],$$

where $\delta = \epsilon - \eta$, so that $E(\delta) = 0$ and $\operatorname{var}(\delta) = 2\sigma^2$, and the double prime superscript denotes a second derivative. Estimating the bias term is not straightforward; Coffin and Sukhatme (1997) suggest using kernel methods to estimate the densities of S'_P, S'_N, from which they derive approximations to the densities of S_P, S_N and hence an expression for the estimated bias that depends only on σ^2 and the data. The variance of the measurement errors can generally be gauged from the substantive application and historical knowledge, so $\widehat{AUC}'$ can readily be corrected for bias. Coffin and Sukhatme (1997) demonstrate the efficacy of this correction via Monte Carlo experiments.

Having considered bias, the next step is to find appropriate adjustments to confidence intervals for AUC in the presence of measurement errors, and a number of authors have considered this aspect. Faraggi (2000) considers the parametric approach using the binormal model. From Section 2.5, we see that for this model

$$AUC = \Phi\left(\frac{\mu_P - \mu_N}{\sqrt{\sigma_P^2 + \sigma_N^2}}\right).$$

Faraggi assumes that $\sigma_P^2 = \sigma_N^2 = \sigma^2$ and also, like Coffin and Sukhatme above, that all measurement errors have a common variance σ_E^2. He then sets $\theta^2 = \sigma_E^2/\sigma^2$, which provides a standardized measure of the size of the measurement errors (and lists references to various studies

4.4. ERRORS IN MEASUREMENTS

that give numerical estimates of θ^2 values in different applications). Under these assumptions, the actual observations S'_{Pi} and S'_{Ni} have variance $\sigma^2(1+\theta^2)$ rather than σ^2, and hence

$$\text{AUC}' = \Phi\left(\frac{\mu_P - \mu_N}{\sqrt{[2\sigma^2(1+\theta^2)]}}\right).$$

A confidence interval can thus be constructed in the usual way starting from AUC' (e.g., by using the method of Owen *et al.*, 1964), and this interval has limits $\Phi(\delta_L), \Phi(\delta_U)$ where δ_L, δ_U are the lower and upper confidence limits for $\mu_P - \mu_N$. Thus, to take measurement errors into account, δ_L and δ_U must both be multiplied by $\sqrt{1+\theta^2}$, and the limits for AUC altered accordingly.

Faraggi shows numerically via Monte Carlo experiments that by not taking measurement errors into account the computed confidence intervals for AUC have actual coverage substantially less than the stated nominal value, so there is a corresponding risk of obtaining seriously misleading conclusions. However, his method requires θ^2 to be known. When it is not known (often the case in practice), he suggests checking how sensitive the computed interval is to variation in the value of θ^2, but this begs the question of what to do if the interval is very unstable over the studied range of values. Reiser (2000) therefore tackles the problem of adjusting the interval when the measurement errors have to be estimated. He considers several scenarios in which such estimation is possible, and focusses on studies that include replication of observations over individuals. He relaxes the assumption of a common variance for observations from populations N and P, and likewise allows different variances for the measurement errors in these two populations. He then uses two different approaches to derive expressions for confidence intervals in the presence of measurement errors; one approach is a generalization of the method of Reiser and Guttman (1986), while the other uses the delta method. A simulation study investigates the performance of these methods, and additionally shows that the practice of averaging over replicates and ignoring measurement error is not to be recommended. Schisterman *et al.* (2001) extend the investigations to the case where data from an external reliability study can be used to estimate the variance of the measurement errors, while Tosteson *et al.* (2005) explore the effects of possibly heterogeneous measurement errors under the binormal model with either known or estimated measurement error variances.

4.5 Further reading

Given the connections between AUC and the Mann-Whitney U-statistic, it is unsurprising that research on sample size determination for this latter statistic impacts upon sample size determination for AUC studies. The recent paper by Shieh et al. (2006) provides a bridge between the two areas. Moreover, the link with $\Pr(Y < X)$ of both these statistics means that there is a rich body of literature outside the classification or ROC area in which interesting results may be found. The reader is referred to Mee (1990), Simonoff et al. (1986), and Reiser and Faraggi (1994), in addition to those references already mentioned in this chapter. Wolfe and Hogg (1971) were early proponents of AUC for measuring the difference between two distributions.

Chapter 5

ROC curves and covariates

5.1 Introduction

Once a classifier $S(\boldsymbol{X})$ has been constructed from the vector $\boldsymbol{X}$ of primary variables and is in use for allocating individuals to one or other of the populations N and P, it frequently transpires that a further variable or set of variables will provide useful classificatory information which will modify the behavior of the classifier in one or both of the populations. Such additional variables are termed *covariates*, and for maximum benefit should clearly be incorporated into any analyses involving the classifier. Let us denote a set of c covariates by the vector $\boldsymbol{Z}$, recognizing that in many practical applications we may have just one covariate Z in which case $c = 1$.

Examples of covariates can arise in various ways. Perhaps the most common one is in radiological or medical situations, where the classifier S is just a single measurement (i.e., a "marker," such as blood glucose level, forced expiratory level, amplitude of auditory signal, and the like) obtained directly using a laboratory instrument, while additional information such as the age and gender of the patient or the severity of their illness may have an effect on the distributions of S. However, other examples also arise when the classifier has been constructed from primary variables $\boldsymbol{X}$ and further variables $\boldsymbol{Z}$ come to light subsequently, either as being of potential importance in distinguishing between P and N, or in providing additional information that enables the performance of $S(\boldsymbol{X})$ to be assessed more accurately. In all such cases, it is necessary to adjust the ROC curve and summaries derived therefrom before proceeding to draw any inferences.

In particular, it will often be relevant to compute the ROC curve and allied summaries at particular values of the covariates Z, in order to relate these *covariate-specific* curves and summaries to sample members having the given covariate values. Ignoring covariate values leads to the calculation of a single *pooled* ROC curve (the "pooling" being over all the possible covariate values), and Pepe (2003, pp. 132-136) provides several important results linking the covariate-specific and the pooled ROC curves. If the covariates do not affect the (true) ROC curve (i.e., if the ROC curve is the same for every set of covariate values) but they *do* affect the distribution of S in population N, then the pooled ROC curve will be attenuated (i.e., it will at best equal the true ROC curve but will otherwise lie below it). On the other hand, if the covariates affect the ROC curve but not the distribution of S in population N, then the pooled ROC curve will be a weighted average of the covariate-specific ROC curves. Such results emphasize the need for taking covariate information into account, particularly if using the analysis to guide decisions about specific individuals.

Covariate adjustment of discriminant or classification functions has quite a long history, major early work being that by Cochran and Bliss (1948), Rao (1949), and Cochran (1964). These papers were primarily concerned with selection of discriminatory variables rather than with the adjustment of the classification rule to allow for the presence of covariates. This latter objective requires the formulation of appropriate probability models, and results from this approach are summarized by Lachenbuch (1977) and McLachlan (1992, pp. 74-78). However, all this work is focussed on the classification rule itself so provides essentially just a background to our own interest here, which assumes that the rule has already been fixed and so concentrates on the adjustment of quantities derived from it.

The first papers in this regard were those by Guttman *et al.* (1988) and Tosteson and Begg (1988). The former obtained $p(X > Y)$ as a function of covariates in the context of stress-strength models, while the latter considered purely ordinal classification scores. Adjusting ROC curves derived from continuous classification scores has been considered by Smith and Thompson (1996), Pepe (1997, 1998, 2000), Faraggi (2003), Janes and Pepe (2006, 2007), Pepe and Cai (2004), and Janes *et al.* (2008). Adjustment of summary values derived from the ROC curve has further been studied by Faraggi (2003) and Dodd and Pepe (2003 a,b). These are the studies that we focus on in the present chapter; we look first at the adjustment of the curve, and then at the adjustment of summaries derived from it.

5.2 Covariate adjustment of the ROC curve

Two distinct approaches can be followed: indirect adjustment, in which the effect of the covariates on the distributions of S is first modeled in the two populations and the ROC curve is then derived from the modified distributions, and direct adjustment in which the effect of the covariates is modeled on the ROC curve itself.

5.2.1 Indirect adjustment

We see from the introductory section that the earliest indirect approach was formulated by Guttman *et al.* (1988). They specifically considered the question of obtaining confidence bounds on the quantity $p(X > Y | z_1, z_2)$, where X and Y are normal variates and z_1, z_2 are explanatory variables. In our terms this corresponds to the estimation of $p(S_P > S_N)$ as a function of covariates and assuming normal distributions of S in both N and P. Given the equivalence of the two quantities it might be more properly dealt with as direct modeling of AUC. However, Faraggi (2003) formulates the Guttman *et al.* model within the ROC setting, so we take this formulation as the natural starting point of the present section.

In its most general form, suppose there is a set of c_P covariates $\mathbf{Z}_P$ associated with population P and a set of c_N covariates $\mathbf{Z}_N$ associated with population N. Of course in most practical applications many if not all of the covariates will be common to both populations, but there is no necessity for the two sets to be identical. Now if we let α_P, α_N be two scalars and $\boldsymbol{\beta}_P, \boldsymbol{\beta}_N$ be c_P- and c_N-element vectors of unknown parameters, then the means of S_P and S_N for given values of the covariates can be modeled as $\mu_P(\mathbf{Z}_P) = \alpha_P + \boldsymbol{\beta}_P^T \mathbf{Z}_P$ and $\mu_N(\mathbf{Z}_N) = \alpha_N + \boldsymbol{\beta}_N^T \mathbf{Z}_N$. The specification is then completed by assuming normal distributions for S_P and S_N, with standard deviations σ_P, σ_N respectively.

We see that this model is essentially the same as the one underlying the binormal development of Section 2.5, the only difference being in the specification of the population means. Hence it follows that the equation of the ROC curve is given by

$$y = \Phi\left(\frac{\mu_P(\mathbf{Z}_P) - \mu_N(\mathbf{Z}_N) + \sigma_N \times \Phi^{-1}(x)}{\sigma_P}\right) \quad (0 \le x \le 1).$$

Ordinary least-squares regression can be used to obtain point estimates of $\alpha_P, \alpha_N, \boldsymbol{\beta}_P, \boldsymbol{\beta}_N, \sigma_P$, and σ_N, and substitution of these estimates into the above formula at given values $\mathbf{z}_P$, $\mathbf{z}_N$ of $\mathbf{Z}_P$, and $\mathbf{Z}_N$ will yield the covariate-specific ROC curves.

Smith and Thompson (1996) followed up the above model with another regression-based approach, but replacing the assumption of normality by Weibull distributions with a common scale parameter in both P and N. Faraggi (2003) pointed out that this assumption was quite restrictive, as the populations differed markedly in scale for many radiological applications, and that while the results could in principle be generalized to the case of unequal scale parameters this could only be done at the cost of considerable mathematical complication. Moreover, several comments have already been made in previous chapters regarding the ease with which such distributional assumptions can fail in practical applications, signalling the need for a model that has less restrictive distributional assumptions. Such considerations led to the approach of Pepe (1998) and its generalization by Zhou et al. (2002).

Pepe (1998) took a continuous-score analogy of Tosteson and Begg's (1988) ordinal-score approach, and modeled S_P, S_N as coming from some arbitrary location-scale family but with means and standard deviations as specified above for the normal model. If the distribution function of the chosen member of the location-scale family is $H(\cdot)$ on standardized scale (i.e., for mean 0 and variance 1), then the equation of the ROC curve is given by

$$y = 1 - H\left(a + bH^{-1}(1-x) + \mathbf{c}^T \mathbf{Z}\right) \quad (0 \le x \le 1),$$

where $a = \alpha_N - \alpha_P, b = \sigma_N/\sigma_P$, and $\mathbf{c} = \left(\frac{1}{\sigma_P}\right)(\boldsymbol{\beta}_N - \boldsymbol{\beta}_P)$.

To obtain estimates of the covariate-specific ROC curves in any practical application, one needs to estimate the parameters $\alpha_P, \alpha_N, \boldsymbol{\beta}_P, \boldsymbol{\beta}_N, \sigma_P$, and σ_N, and Pepe (1998) notes that parameter estimates may be obtained without specifying a form for $H(\cdot)$ by using quasi-likelihood methods (McCullagh and Nelder, 1989). Suppose therefore that we have a sample of n individuals, some of whom are from population P and some from population N. Let $p(i)$ denote the population membership of individual i (so that $p(i) = $ P or $p(i) = $ N), s_i be the classification

5.2. COVARIATE ADJUSTMENT OF THE ROC CURVE

score, and $\mathbf{z}_{p(i)}$ be the set of covariate values for individual i. Then $\mu_{p(i)}(\mathbf{z}_{p(i)}) = \alpha_{p(i)} + \boldsymbol{\beta}_{p(i)}^T \mathbf{z}_{p(i)}$ is the mean of individual i as a function of the regression parameters, and $\sigma_{p(i)}$ is the standard deviation parameter for individual i. Let $\boldsymbol{\theta}$ denote the collection of all k regression parameters $\alpha_P, \alpha_N, \boldsymbol{\beta}_P, \boldsymbol{\beta}_N$, with θ_j denoting any one particular parameter from this set. Then the estimators $\hat{\theta}_j$ of the mean parameters are chosen to satisfy

$$\sum_{i=1}^n \left(\frac{\partial \mu_{p(i)}(\mathbf{z}_{p(i)})}{\partial \theta_j} \right) \left(\frac{s_i - \mu_{p(i)}(\mathbf{z}_{p(i)})}{\sigma_{p(i)}} \right) = 0 \quad \text{for } j = 1, \ldots, k,$$

and the variance estimators are

$$\hat{\sigma}_p^2 = \left(\sum_{i=1}^n I[p(i) = p]\{s_i - \mu_{p(i)}(\mathbf{z}_{p(i)})\}^2 \right) \div \left(\sum_{i=1}^n I[p(i) = p] \right)$$

for $p = $ P or N, using the indicator function $I(x = a)$ which takes value 1 if $x = a$ and 0 otherwise. These are all simultaneous equations, but Pepe (1998) points out that given initial values for all the unknown parameters one can iterate between solving for the regression parameters and updating the variance estimates until convergence is achieved for all values. The baseline distribution function $H(y)$ can then be estimated nonparametrically as the number of standardized residuals $\{s_i - \mu_{p(i)}(\mathbf{z}_{p(i)})\}/\hat{\sigma}_{p(i)}$ that are less than or equal to y, divided by n.

Zhou et al. (2002) proposed an extended version of the above model and procedure, in which the standard deviations σ_P and σ_N as well as the means $\boldsymbol{\mu}_P$ and $\boldsymbol{\mu}_N$ are allowed to be functions of the covariates. This extends the number of estimating equations in the iterative process, but it essentially remains a two-stage process with the mean and standard deviation parameters being estimated in the first stage and the distribution function $H(y)$ being obtained nonparametrically from standardized residuals in the second. As an alternative to methods based on an arbitrary location-scale model, Heagerty and Pepe (1999) and Zheng and Heagerty (2004) detail a method based on specific semiparametric location-scale models, using quasi-likelihood to estimate regression parameters and empirical distribution functions of fitted residuals to estimate survival functions.

5.2.2 Direct adjustment

Instead of separately modeling the effects of the covariates on the two classification score distributions and then deriving the induced covariate-specific ROC curve from the modified distributions, we can alternatively model the effects of the covariates directly on the ROC curve itself. Among other potential advantages, such an approach means that any parameters associated with the covariates have a direct interpretation in terms of the curve. At the heart of this approach is the specification of an appropriate model that will capture the effects of the covariates on the ROC curve in a flexible and interpretable manner. In view of the familiarity of least-squares regression methods for adjusting measured variates such as the classification score S, a linear regression-like relationship of covariates to ROC curve is desirable. However, there is also a need for considerable flexibility in terms of the way this relationship acts on the curve while preserving the constraints that both domain and range of the curve are in the interval (0, 1) and that the curve is monotonically increasing over this interval.

Such flexibility and constraint requirements have led to the use of *generalized linear model* (GLM; McCullagh and Nelder, 1989) methodology for direct modeling of ROC curves. Specifically, the popular approach is via the ROC-GLM model which posits the equation of the curve to be of the form

$$h(y) = b(x) + \boldsymbol{\beta}^T \boldsymbol{Z},$$

where $b(\cdot)$ is an unknown baseline function monotonic on (0, 1) and $h(\cdot)$ is the link function, specified as part of the model and also monotonic on (0, 1). $\boldsymbol{Z}$ is as usual the vector of covariates (but now a single vector rather than two separate population-specific vectors) and $\boldsymbol{\beta}$ are the regression parameters associated with the covariates. Various possible link functions can be employed, the most familiar ones from GLM theory being the probit with $h(y) = \Phi^{-1}(y)$, the logistic with $h(y) = \log(y/[1-y]) = \text{logit}(y)$, and the logarithmic with $h(y) = \log(y)$.

In fact, Pepe (2003, p154) shows that if the classification scores have means $\mu_P(\boldsymbol{Z}) = \alpha_P + \boldsymbol{\beta}_P^T \boldsymbol{Z}$, $\mu_N(\boldsymbol{Z}) = \alpha_N + \boldsymbol{\beta}_N^T \boldsymbol{Z}$ and standard deviations σ_P, σ_N in the two populations respectively, then the induced ROC curve is a member of the ROC-GLM family. However, the reverse is not true in that a ROC-GLM form of curve does not necessarily imply the existence of a corresponding model for the classification scores. Hence there is greater flexibility implicit in such direct modeling of the curve. Pepe (2003) also lists some other advantages and disadvantages of direct modeling vis-à-vis the separate modeling of classification score

5.2. COVARIATE ADJUSTMENT OF THE ROC CURVE

distributions. Direct modeling allows the inclusion of interactions between covariates and x values, the modeling of accuracy parameters, and the comparison of different classifiers. On the other hand, the modeling of variates such as classification scores is a familiar statistical process whose methods are generally statistically efficient, whereas little is at present known about the efficiency of methods for direct modeling of ROC curves. Nonetheless, we now consider the fitting of ROC-GLM models to data.

First recollect that at the end of Section 3.3.4 on binary regression methods we said that further methods had been developed by Pepe and coworkers, but that since they allowed for the incorporation of covariates their description would be delayed until the present chapter. Accordingly, we now consider them. The description that follows is a synthesis of the work by Pepe (1998, 2000, 2003), Alonzo and Pepe (2002), Cai and Pepe (2003), and Pepe and Cai (2004), and makes use of the concept of placement values mentioned in Section 3.5.1. For a more comprehensive discussion of placement values see Pepe and Cai (2004); here we just use their necessary features for the estimation process.

Given the classification score S and its cumulative distribution functions F, G in populations N, P respectively, we typically use N as the reference population and define the placement value of S by $U = 1 - F(S)$. It is the proportion of population N with values greater than S. Standard theory thus shows that the distribution of placement values U_N in population N is uniform on (0, 1), while the distribution of placement values U_P in population P quantifies the separation between the two populations — for well separated populations the S_P scores are all higher than the S_N scores, so that the U_P values will all be very small, but as the populations progressively overlap then the U_P and U_N values will intermingle to a greater and greater extent. In fact, Pepe and Cai (2004) show that $\text{ROC}(t) = p(U_P \leq t)$, i.e., the ROC curve is just the cumulative distribution function of the U_P values. Thus if we now envisage a set of independent observations $S_{P1}, S_{P2}, \ldots, S_{Pn}$ from population P, and we use the notation we established in Section 3.5.1, the placement value of the ith of these observations is S_{Pi}^N. Using the indicator function $I(\cdot)$ defined above, $U_{it} = I(S_{Pi}^N \leq t)$ is the binary random variable indicating whether or not the ith placement value is less than or equal to t, so it follows directly that $E(U_{it}) = \text{ROC}(t)$.

Now return to the ROC-GLM model. Let us first consider the

case of no covariates, and suppose that the baseline function $b(\cdot)$ has the form $\sum_k \alpha_k b_k(\cdot)$ for some specified functions $b_1, b_2, \ldots, b_K$. Thus the ROC-GLM model is $h(y) = \sum_k \alpha_k b_k(x)$. For example, if $h(\cdot) = \Phi^{-1}(\cdot), b_1(x) = 1$, and $b_2(x) = z_x = \Phi^{-1}(x)$, then the ROC-GLM model is precisely the binormal model as given in Section 2.5. So if we now have samples of scores s_{Pi} and s_{Ni} from the two populations then we can use standard GLM methods to estimate the parameters of the model, and Pepe (2003) sets out the process. First choose a set of values of t over which the model is to be fitted. Then for each t obtain the binary placement value indicators $u_{it} = I(s_{Pi}^N \leq t)$ for $i = 1, \ldots, n_P$. Finally, binary regression with link function $h(\cdot)$ and covariates $b_1(t), b_2(t), \ldots, b_K(t)$ will provide estimates of the model parameters $\alpha_1, \alpha_2, \ldots, \alpha_K$. Note that the t values are false positive fractions for the purpose of ROC analysis, so we can choose up to $n_N - 1$ values in general (with fewer if there are ties in the s_N values). However, Pepe (2003) argues that the full set might generally produce too large and unwieldy a set of data for the fitting process, and Alonzo and Pepe (2002) show that a relatively small set can be chosen without undue loss of efficiency.

If we now wish to incorporate covariates, the general idea and procedure is the same but there are some complicating features. Assume again that the baseline function is fully parameterized as $b(t) = \sum_k \alpha_k b_k(t)$ for specified functions $b_1, b_2, \ldots, b_K$. We thus need to calculate the set of placement values in order to start the analysis, but note now that the score distributions in the two populations will depend on the covariates, and in particular the placement values will depend on those covariates that *are common to both populations*. So let us write $\mathbf{Z} = (\mathbf{Z}_c, \mathbf{Z}_p)$ where $\mathbf{Z}_c$ are the common covariates and $\mathbf{Z}_p$ are the ones specific to population P. Thus the placement value of S is now defined by $U = 1 - F(S \mid \mathbf{Z}_c)$, and the *covariate-specific* ROC curves are now the appropriate *conditional* distributions: $\mathrm{ROC}(t \mid \mathbf{Z}) = p(U_P \leq t \mid \mathbf{Z})$. However, the binary random variable $U_{it} = I(S_{Pi}^N \leq t \mid \mathbf{Z}_c)$ continues to follow a generalized linear model, with the specific form

$$h(y) = b(x) + \beta^T \mathbf{Z}.$$

When it comes to fitting the model to data, we again have to first select a set of values of t, but the introduction of covariates brings a further complication. Previously, the placement values could be obtained directly using empirical proportions of the sample data. Now, however, the dependence of the placement values on the values of the covariates $\mathbf{Z}_c$ means that placement values have to be *estimated* by first estimating the conditional distribution function F of scores given $\mathbf{Z}_c = \mathbf{z}_c$ in population N. This estimation can be done in any of a number of ways using the sample classification scores from population N (see, for example, Pepe, 2003, Section 6.2). Having estimated this distribution function, placement values can be obtained and hence for each selected t the binary placement value indicators $u_{it} = I(s_{Pi}^N \leq t)$ for $i = 1, \ldots, n_P$ and the baseline function values $b_1(t), b_2(t), \ldots, b_K(t)$ can be calculated. These sets of values plus the corresponding values of the covariates $\mathbf{Z}_c$, $\mathbf{Z}_p$ form the data, and parameters are estimated by fitting the marginal generalized linear binary regression model.

As a final comment, we can note that if a parameterized baseline function $b(\cdot)$ is better left unspecified, then a semiparametric ROC-GLM can be fitted. This requires an extra iterative scheme to be incorporated within the above structure, for estimating the baseline function nonparametrically. See Cai and Pepe (2003) for details.

5.2.3 Applications

Pepe (1998 and 2003) has applied the methods of this section to a hearing test study described by Stover *et al.* (1996), so we here summarize the results of these analyses. In the study, 107 hearing-impaired subjects (the sample from population P) and 103 normally-hearing subjects (the sample from population N) underwent the distortion product otoacoustic emissions (DPOAE) test for hearing impairment. The test result (i.e., the classification score or marker S) was the negative signal-to-noise ratio at a given frequency and input stimulus level, oriented in this direction to ensure that higher values indicated greater hearing impairment (population P). An independent gold-standard behavioral test yields a hearing threshold in decibels (dB), and populations P and N are defined according as the hearing thresholds of their subjects exceed or do not exceed 20 dB.

The DPOAE test was applied nine times to one ear of each subject, for all combinations of 3 levels of stimulus frequency f and 3 levels of stimulus intensity i. Stimulus frequency and stimulus intensity thus comprised two covariates, and these were used on their original scales $Z_f = f/100$ (Hz) and $Z_i = i/10$ (dB) respectively. Additionally, it was felt that the degree of hearing loss might affect S, so this was measured by a third covariate $Z_d = $ (hearing threshold - 20)/10 (dB) which takes values greater than 0 in the hearing-impaired group but is absent in the normally-hearing group.

First taking the indirect adjustment approach of Section 5.2.1, the parameters of the model are $\mu_P = \alpha_P + \beta_{Pf}Z_f + \beta_{Pi}Z_i + \beta_{Pd}Z_d, \mu_N = \alpha_N + \beta_{Nf}Z_f + \beta_{Ni}Z_i, \sigma_P$ and σ_N. Assuming normality of classification scores, ordinary least squares (i.e., maximum likelihood) estimates of these parameters were $24.59, 0.33, -5.77, 3.04, 1.11, -0.14, -0.86, 8.01$, and 7.74 respectively, thus leading to the induced covariate-specific ROC curve equation

$$y = \Phi(2.93 + 0.06Z_f - 0.61Z_i + 0.38Z_d + 0.97\Phi^{-1}[x]).$$

When the assumption of normality was relaxed, Pepe (2003) reported that the quasi-likelihood estimates of the parameters, obtained from the estimating equations in Section 5.2.1, were the same as the previous maximum likelihood ones, and the nonparametrically estimated baseline distribution function $H(y)$ was very close to the normal one. In all cases, the (analytical asymptotic) standard errors were sufficiently small for the parameter estimates to be deemed significant at the 5% level. These results therefore suggest that the DPOAE test is more accurate at higher values of Z_f, less accurate at higher values of Z_i, and better able to discriminate between the two populations the higher the value of Z_d.

If the direct adjustment methods of Section 5.2.2 are applied to the ROC-GLM model with probit link function, the form of the fitted ROC curve is as above but the coefficients of Z_f, Z_i, and Z_d are now $0.05, -0.41$, and 0.44 respectively. The qualitative conclusions are thus the same as before. However, standard errors must now be obtained by bootstrapping, and while the coefficients of Z_i and Z_d remain significant at 5% the coefficient of Z_f is no longer significant. The conclusion is therefore less clear-cut than before.

As a second example, we show that the regression framework developed above can be used to compare two ROC curves. Wieand *et al.* (1989) reported on the use of two monoclonal antibodies (CA-125 and CA19-9) as potential classifiers of pancreatic cancer. Both antibodies were measured in the sera of $n_P = 90$ patients with the disease and $n_N = 51$ control subjects without it. Pepe (2000) shows the empirical ROC curves for both potential classifiers.

If we now define a covariate Z as taking value 0 for antibody CA-125 and value 1 for antibody CA19-9, and assume normality of both classifiers, then the single model

$$y = \Phi(\alpha_1 + \beta_1 \Phi^{-1}[x] + \alpha_2 Z + \beta_2 Z \Phi^{-1}[x])$$

is a representation of the binormal model ROC curves of both classifiers. Parameters α_1, β_1 relate to classifier CA-125, while α_2, β_2 indicate how the parameters for CA19-9 differ from those for CA-125. Using the methods of Section 5.2.2 with x restricted to lie in $(0, 0.2)$ and an empirically estimated baseline function, Pepe (2000) obtained the estimated CA-125 ROC curve as $\Phi(0.91 + 1.35\Phi^{-1}[x])$ with $\hat{\alpha}_2 = 0.23$ and $\hat{\beta}_2 = -0.91$ as the adjustments needed to form the ROC curve for CA19-9. The small value of $\hat{\alpha}_2$ indicates little difference in the two ROC intercepts, but the large negative value of $\hat{\beta}_2$ shows that in the given region $x \in (0, 0.2)$ (where $\Phi^{-1}(x)$ is negative) the curve for CA19-9 clearly lies above that for CA-125. Moreover, using bootstrapping to obtain standard errors and the variance-covariance matrix of the coefficients, a formal test of hypothesis clearly rejects the null hypothesis of equality of ROC curves.

Some further comments on the use of ROC methodology for comparing curves are given in the next chapter.

5.3 Covariate adjustment of summary statistics

In many applications the chief interest is in summary statistics of the ROC curve rather than in the curve itself, and so we may need to adjust such summary measures for the values of any covariates. Nearly all theoretical attention has been concentrated on AUC and partial AUC, with very little accorded to other potential ROC summaries.

5.3.1 Adjustment of AUC

If the ROC curve has been adjusted indirectly, by first modeling the effects of the covariates on the classification score S in each population, then the effects on AUC are obtained easily on expressing AUC in terms of the model parameters and substituting estimates of the parameters in this expression.

Thus, for example, in the case of the Guttman *et al.* (1988) and Faraggi (2003) models described in Section 5.2.1 above, the binormal form immediately leads to the following expression for AUC at values z_P, z_N of the covariates $\mathbf{Z}_P$ and $\mathbf{Z}_N$:

$$\mathrm{AUC}(z_P, z_N) = \Phi[\delta(z_P, z_N)],$$

where $\delta(z_P, z_N) = \frac{\mu_P(z_P) - \mu_N(z_N)}{\sqrt{\sigma_P^2 + \sigma_N^2}}$ for $\mu_P(z_P) = \alpha_P + \beta_P^T z_P$ and $\mu_N(z_N) = \alpha_N + \beta_N^T z_N$. Estimating all the parameters by ordinary least squares, as for the ROC curve estimation, and substituting into these expressions thus yields $\hat{\delta}(z_P, z_N)$ and $\widehat{\mathrm{AUC}}(z_P, z_N)$.

Moreover, standard asymptotic normal arguments as applied to regression models provide approximate confidence intervals. If we write $\tilde{\mathbf{Z}}_P, \tilde{\mathbf{Z}}_N$ for the matrices of covariate values of sample members from P and N, then an approximate $100(1 - \alpha)\%$ confidence interval for $\delta(z_P, z_N)$ is given by

$$\hat{\delta}(z_P, z_N) \mp \sqrt{\frac{1}{\hat{M}(z_P, z_N)} + \frac{\hat{\delta}^2(z_P, z_N)}{2\hat{f}}} \, \Phi^{-1}\left(1 - \frac{\alpha}{2}\right)$$

where $\hat{M}(z_P, z_N) = \frac{\hat{\sigma}_P^2 + \hat{\sigma}_N^2}{a_P^2 \hat{\sigma}_P^2 + a_N^2 \hat{\sigma}_N^2}$ for $a_i^2 = z_i^T (\tilde{\mathbf{Z}}_i^T \tilde{\mathbf{Z}}_i)^{-1} z_i$ ($i = $ P, N) and $\hat{f} = \frac{\sigma_P^2 + \sigma_N^2}{\sigma_P^4/(n_P - c_P) + \sigma_N^4/(n_N - c_N)}$. Thus, if v_L and v_U are the values at the lower and upper ends respectively of this interval, then an approximate $100(1 - \alpha)\%$ confidence interval for $\widehat{\mathrm{AUC}}$ is $(\Phi[v_L], \Phi[v_U])$.

Of course, modeling the classification scores in terms of covariates as above will generally require fairly strong or specific assumptions. Moreover, situations may arise in which the classification scores are not available but AUC values are, for different settings of the covariates. To open up analysis in the latter case, and to relax the necessary assumptions in the former, several authors have considered direct regression modeling of AUC.

The first approach was by Thompson and Zucchini (1989), in the case where replicate AUC values $\widehat{\mathrm{AUC}}_{ki}$ are available at each of K distinct covariate settings z_k ($k = 1, \ldots, K$). The basic regression model is

$$E(h[\widehat{\mathrm{AUC}}_k]) = \alpha_k + \beta_k^T z_k,$$

where $h(\cdot)$ is a known strictly monotonic transformation from $(0, 1)$ to $(-\infty, \infty)$. This transformation is needed if the linear predictor is not to be restricted, while the AUC values are constrained to lie in the interval $(0, 1)$. Familiar transformations such as the logit or probit are the natural ones to use in this context. Pepe (1998) has discussed this model, using as the estimator of AUC_k the Mann-Whitney U-statistic (Section 3.5.1) on the individuals whose covariate values are z_k and including the estimation of the variance of $h[\widehat{\text{AUC}}_k]$. However, Dodd and Pepe (2003a) pointed out two major weaknesses of the method, namely that continuous covariates cannot be modeled, and that the basic regression assumption of equal variances may not be satisfied since there will generally be different numbers of individuals at the different covariate settings.

Accordingly, Dodd and Pepe (2003a) extended the model to allow all types of covariates, and proposed an estimation method based on binary regression. As in Section 5.2.1, we suppose that there is a set of c_P covariates $\boldsymbol{Z}_P$ associated with population P and a set of c_N covariates $\boldsymbol{Z}_N$ associated with population N, and write $\boldsymbol{Z}$ for the aggregation of all covariates. The AUC regression model is then

$$E(h[\widehat{\text{AUC}}]) = \alpha + \boldsymbol{\beta}^T \boldsymbol{z}$$

for parameters α, $\boldsymbol{\beta}$ and monotone increasing link function $h(\cdot)$, with the probit and logit again the natural link functions. To estimate the parameters define $U_{ij} = I(S_{Pi} > S_{Nj})$, where S_{Pi} and S_{Nj} are the classification scores for the ith individual from P and the jth individual from N respectively, and $I(\cdot)$ is the indicator function defined in Section 5.2.1 above. Then, since $E(U_{ij} \mid \boldsymbol{Z}) = \Pr(S_{Pi} > S_{Nj} \mid \boldsymbol{Z}) = \text{AUC} \mid \boldsymbol{Z}$, the AUC regression model is a generalized linear model for the binary variables U_{ij}. Standard methods based on estimating functions can be used, and Dodd and Pepe (2003a) give full details of their implementation. However, since the same individual i from P contributes to each U_{ij} for all j (and likewise the same individual j from N contributes to each U_{ij} for all i), all the U_{ij} are cross-correlated so the standard asymptotic theory for generalized linear models is no longer appropriate. Dodd and Pepe (2003a) provide the necessary amendments to the theory, and show using simulation studies that the method produces estimates with small bias and reasonable coverage probability.

5.3.2 Adjustment of partial AUC

Following their work on partial AUC estimation, as described in Section 3.5.2, Dodd and Pepe (2003b) extended the binary regression model used there to allow for covariate modeling of the partial AUC. Maintaining our notation of 3.5.2, the covariate-specific partial AUC becomes $\text{PAUC}(f_1, f_2 \mid \mathbf{Z}) = \Pr(S_P > S_N, f_1 \leq 1 - F(S_N) \leq f_2 \mid \mathbf{Z})$. For a specified link function $h(\cdot)$, the model is then given by

$$E(h[\widehat{\text{PAUC}}(f_1, f_2 \mid \mathbf{Z})]) = \alpha + \boldsymbol{\beta}^T \mathbf{z}.$$

Logit and probit are again the natural link functions, but since $\text{PAUC}(f_1, f_2)$ has an upper bound of $f_2 - f_1$ an adjustment of the form $h(u) = \log(u/[f_2 - f_1 - u])$ is appropriate.

To estimate the parameters of the model, we again set up a binary regression situation of the same form as for the AUC above but amending the quantities U_{ij} in order to accord with the constraints imposed in the partial AUC - viz $\text{PAUC}(f_1, f_2) = p(S_P > S_N, f_1 \leq 1 - F(S_N) \leq f_2)$. We thus need to consider all possible pairs of observations, one from each population, that satisfy the required constraint and for which the observation from P exceeds the one from N. Recollect from Section 3.5.2 that pairs (S_{Ni}, S_{Pj}) satisfy the constraint if the rank r_j of S_{Pj} among the sample from P, i.e., $r_j = \sum_{k=1}^{n_P} I(S_{Pk} \leq S_{Pj})$, is such that $f_1 \leq \frac{r_j}{n_P} \leq f_2$. Thus instead of the U_{ij}, we need to use $V_{ij} = I(s_{Pj} > s_{Ni})I(f_1 \leq \frac{r_j}{n_P} \leq f_2)$.

Now conditioning on $\mathbf{Z}$, it follows that $E(V_{ij} \mid \mathbf{Z}) = \Pr(S_{Pi} > S_{Nj}, nf_1 \leq 1 - F(S_N) \leq f_2 \mid \mathbf{Z}) = \text{PAUC}(f_1, f_2 \mid \mathbf{Z})$ so that standard binary regression methods on the V_{ij} can be used to estimate the model parameters. Dodd and Pepe (2003b) give the estimating equations, and show that the same asymptotic theory as in Dodd and Pepe (2003a) is appropriate in view of the cross-correlations between the V_{ij}.

5.3.3 Other summary statistics

The only other summary index to have received attention in the context of covariate adjustment is the Youden Index (defined in Section 2.4.3) along with the critical classifier threshold t^* corresponding to it. These quantities were considered by Faraggi (2003) in his formulation of the Guttman et al. (1988) model as given at the start of Section 5.2.1 above. In terms of this model the Youden Index YI is obtained by maximizing over t the expression

$$tp(\mathbf{Z}_P, \mathbf{Z}_N) + tn(\mathbf{Z}_P, \mathbf{Z}_N) - 1 = \Phi\left(\frac{\mu_P(\mathbf{Z}_P) - t}{\sigma_P}\right) + \Phi\left(\frac{t - \mu_N(\mathbf{Z}_N)}{\sigma_N}\right) - 1,$$

and t^* is the value of t yielding the maximum.

5.3 COVARIATE ADJUSTMENT OF ROC SUMMARIES

Faraggi (2003) showed that this maximization yields

$$kt^* = -[\sigma_P^2 \mu_N(\mathbf{Z}_N) - \sigma_N^2 \mu_P(\mathbf{Z}_P)]$$
$$+ \sigma_P \sigma_N \sqrt{[\mu_N(\mathbf{Z}_N) - \mu_P(\mathbf{Z}_P)]^2 + (\sigma_N^2 - \sigma_P^2) \log(\sigma_N^2/\sigma_P^2)},$$

where $k = (\sigma_N^2 - \sigma_P^2)$, and hence that

$$YI(\mathbf{Z}_P, \mathbf{Z}_N) = \Phi\left(\frac{\mu_P(\mathbf{Z}_P) - t^*}{\sigma_P}\right) + \Phi\left(\frac{t^* - \mu_N(\mathbf{Z}_N)}{\sigma_N}\right) - 1.$$

For given values $\mathbf{Z}_P = \mathbf{z}_P$, $\mathbf{Z}_N = \mathbf{z}_N$, parameter estimates are obtained from the mean regressions as before and substituted into the above expressions to yield point estimates $\widehat{YI}, \hat{t}^*$. Faraggi (2003) suggests that pointwise confidence intervals for YI and t^* can then be obtained as functions of $\mathbf{z}_P$, $\mathbf{z}_N$ by bootstrapping $\widehat{YI}$ and $\hat{t}^*$.

5.3.4 Applications

Dodd and Pepe (2003a) demonstrated the application of their method for modeling the effects of covariates on AUC, as outlined in Section 5.3.1 above, on the DPOAE hearing-test data described in Section 5.2.3 [although in this paper there were 105 hearing-impaired subjects, rather than the 107 quoted by Pepe (1998)]. They chose the logit transformation for $h(\cdot)$, so that the model is

$$\log\{AUC/(1 - AUC)\} = \beta_0 + \beta_f Z_f + \beta_i Z_i + \beta_d Z_d,$$

where Z_f, Z_i, Z_d are as defined in Section 5.2.3. Standard error estimates were obtained by bootstrapping subjects, and confidence intervals were constructed by assuming normal populations. The model estimates indicated that the AUC log-odds: (i) decreased by 42% for every unit increase in Z_i ($\hat{\beta}_i = 0.58$, with 95% confidence interval $0.43 \leq \beta_i \leq 0.79$); (ii) increased by 85% for every unit increase in Z_d ($\hat{\beta}_d = 1.85$, with 95% confidence interval $1.49 \leq \beta_d \leq 2.50$); and (iii) increased by 7% for every unit increase in Z_f ($\hat{\beta}_f = 1.07$, with 95% confidence interval $0.99 \leq \beta_f \leq 1.16$). The first two coefficients were statistically significant (i.e., significantly different from 1.0) but the third was not. Thus one can conclude that greater accuracy is achievable by using stimuli with lower intensities, that severity of impairment is an important determinant of accuracy, and that more data need to be analyzed before any effect of frequency can be established. These conclusions reinforce those already reached in the previous analyses.

To illustrate the method for detecting the effects of covariates on PAUC, Dodd and Pepe (2003b) analyzed data on the use of prostate-specific antigen (PSA) as a predictor of prostate cancer diagnosis. Serum samples of 240 patients diagnosed with prostate cancer were retrospectively evaluated for (prediagnosis) PSA levels, and compared with PSA levels in age-matched serum samples of 237 subjects who were free of prostate cancer. One potential covariate that could affect accuracy of prediction is the time from serum sampling to diagnosis, and Z_t denoted the number of years prior to diagnosis at which the PSA was measured (e.g., -3 indicated 3 years before diagnosis). Another potential covariate is the method of quantifying PSA, and two possible methods were considered: the total PSA in serum, and the ratio of free to total PSA in serum. The covariate Z_m took value 1 for "total PSA" and 0 for "ratio PSA." False-positive rates reported in the literature for related studies ranged between 0.1 and 0.7, so PAUC(0.0, 0.4) covered the range from 0 to the mid-point of reported values and was thus considered to be a reasonable summary measure to model. Furthermore, the possibility that the effect of time to diagnosis might differ between the two quantification methods was covered by including an interaction term in the model. Adjusting the form of the logit link function $h(\cdot)$ as outlined in Section 5.3.2, and writing a for PAUC(0.0, 0.4), the model fitted to the data was thus

$$\log\{a/(0.4 - a)\} = \beta_0 + \beta_t Z_t + \beta_m Z_m + \beta_i Z_t * Z_m.$$

Standard errors were estimated by bootstrap sampling (200 replicates, using case-control sampling with subjects as the primary units). Parameter estimates, with standard errors in parentheses, were: $\hat{\beta}_0 = 0.99(0.29), \hat{\beta}_t = 0.05(0.05), \hat{\beta}_m = 0.98(0.54)$, and $\hat{\beta}_i = 0.16(0.08)$. Thus we can conclude that total PSA appears to be the better marker for prostate cancer, that PSA accuracy improves (not surprisingly) the closer are the subjects measured to clinical diagnosis of prostate cancer, and that this time effect differs between the two PSA measures. At time = 0, the PAUC log-odds is 2.67 when total PSA is the quantification method.

5.4 Incremental value

In a recent working paper, Janes et al. (2008) suggest that an alternative way of utilizing covariates Z is to treat them as "baseline" predic-

5.4. INCREMENTAL VALUE

tors and to assess the improvement in classification accuracy achieved when S is added to them. This is the *incremental value* of $\mathbf{Z}$, which is quantified by comparing the ROC curve for the combined classifier based on both S and $\mathbf{Z}$ with the one based on $\mathbf{Z}$ alone. Thus here we are looking at the difference between the *joint* effect of S, $\mathbf{Z}$, and the *marginal* effect of $\mathbf{Z}$ in discriminating between populations N and P, whereas in all the other methods in this chapter we have been concerned with the classification accuracy of S *conditional* on $\mathbf{Z}$. We may note that the proposed approach is akin to the familiar "extra sum of squares" principle in regression analysis. If S is a potential explanatory variable in a linear model, while $\mathbf{Z}$ are again covariates, then the worth of S as a predictor is assessed by the difference in regression sums of squares between the model including both S and $\mathbf{Z}$ and the model including just $\mathbf{Z}$.

In order to implement this approach, it is necessary to have a mechanism for combining several different classifiers in a single statistic, and this was provided by McIntosh and Pepe (2002). They drew on the connection between the ROC curve and the Neyman-Pearson Lemma outlined in Section 2.3.2, where it was shown that for a fixed *fp* rate the *tp* rate will be maximized by a classifier whose set of score values S for allocation to population P is given by

$$\mathcal{L}(s) = \frac{p(s|\text{P})}{p(s|\text{N})} \geq k,$$

with k determined by the target *fp* rate. However, by simple application of Bayes' rule, McIntosh and Pepe (2002) showed that the risk score $p(\text{P}|s)$ is a monotone increasing function of the likelihood ratio $\mathcal{L}(s)$, so the optimal classifier is given by $p(\text{P}|s) \geq k^*$ where $k^* = kq/(kq+1)$ for prior odds ratio $q = p(\text{P})/p(\text{N})$. Clearly, therefore, the optimal combination of S and $\mathbf{Z}$ is given by the risk score $p(\text{P}|s,\mathbf{z})$.

Janes *et al.* (2008) mention that a wide variety of binary regression techniques can be used to estimate the risk score. Their preferred choice is logistic regression, in which $p(\text{P}|s, z_1, z_2, \ldots, z_c)$ is modeled as

$$\exp\{f(s, z_1, \ldots, z_c)/[1 + f(s, z_1, \ldots, z_c)]\},$$

where $f(s, z_1, \ldots, z_c)$ is some appropriate function of the classification score s and covariates $z_1, \ldots, z_c$. The nature of the function $f(\cdot)$ is chosen with regard to the problem in hand; for example it could be a simple linear function, or it could contain quadratic and interaction terms, or it could be some more complicated combination of the variables. Whatever the choice, the procedure is carried out by fitting separate logistic regressions to estimate $p(\mathrm{P}|s, z_1, z_2, \ldots, z_c)$ and $p(\mathrm{P}|z_1, z_2, \ldots, z_c)$, calculating the associated predicted values for all individuals in the data set (on the logit-transformed, i.e., linear predictor scale for convenience), and then plotting and comparing the ROC curves for the two models. Software is available for conducting these steps (see the DABS Center web site given in the Appendix).

5.5 Matching in case-control studies

A major area for application of ROC methodology is disease screening and diagnosis, where case-control studies are popular and where matching is a frequent design strategy. Cases are randomly sampled, while controls are matched to cases with respect to covariates such as patient characteristics, disease risk factors, features of the test or test operator, and so on. Matching has been extensively studied in the context of epidemiological associations, but Janes and Pepe (2008) comment that in classification studies the presence of matching is commonly ignored. They therefore go on to consider the various implications.

As a starting point they use a hypothetical data set to show that a crude ROC curve, in which the samples from each population are not adjusted in any way, differs from one that has had covariate adjustment made on it. Their illustrative data relate to a binary covariate and they show that, relative to a covariate-specific ROC curve, the unadjusted ROC curve from an unmatched study is over-optimistic while that from a matched study is attenuated (i.e., downwardly biased). Thus, despite the fact that the samples have been matched, some form of adjustment is still required. Comparison of asymptotic efficiencies with the crude curve shows that matching always increases the efficiency of the nonparametric covariate-adjusted ROC curve estimated at a fixed fp, and an expression is derived for the optimal matching ratio. Interpretation of the covariate-adjusted ROC curve is considered for the case where the covariates Z affect discrimination, and optimal matching ratios are considered in this setting also. However, despite these positive features regarding efficiencies, it is noted that matching with respect to covariates prevents direct estimation of the incremental value. This is because the sampling distribution of the covariates is the same in both populations by dint of the matching. Thus Z is useless as a classification tool by design, and its artificial sampling distribution in the matched data precludes estimation of either risk score or ROC curve.

The conclusion is therefore that matching needs to be considered carefully when designing studies of classification accuracy. It provides considerable benefits in terms of efficiency, but precludes direct estimation of the incremental value of the classifier over the matching covariates. The questions that the investigator wishes to answer are of prime importance when deciding whether or not to match.

5.6 Further reading

A unified approach to the handling of covariates in ROC analysis has been given in the recent series of working papers by Janes and Pepe (2006, 2007) and Janes *et al.* (2008). All the methods of this chapter are based on specific model assumptions, but little attention has been paid as yet to techniques for checking the goodness of fit of these models to data. Such checks are important prior to any analysis, to ensure that it does not produce misleading results, but the only work to date in this area is that by Cai and Zheng (2007). These authors focus on graphical and numerical methods for assessing the goodness of fit of various ROC regression models, using procedures based on cumulative residuals. Asymptotic null distributions are derived, and resampling methods are used to approximate these distributions in practice.

The results of this chapter afford generalizations in several ways. First, if we have more than one classifier in a particular application then we might seek to combine them in an optimal way. One approach is the risk score outlined in Section 5.4, but another one is to look for the linear combination of the different classifiers that maximizes the area under the ROC curve, among the areas under the ROC curves of all possible linear combinations. Schisterman *et al.* (2004) discuss covariate effects on such a linear combination assuming that the separate classifiers, possibly after transformation, follow a multivariate normal distribution. They show how the ROC curve of the linear combination is estimated from the classifiers adjusted for the covariates, and derive approximate confidence intervals for the corresponding AUC.

Second, the connections between AUC and the Mann-Whitney test enable the above methods for ROC regression analysis to be used to extend the Mann-Whitney test so that it can accommodate covariate adjustment. Brumback *et al.* (2006) provide this nonparametric extension, in a fashion that parallels the parametric extension of the t-test to linear models and thereby fills a gap in general nonparametric methodology.

Chapter 6

Comparing ROC curves

6.1 Introduction

So far our discussion has been restricted to studying the properties of single ROC curves, or families of curves determined by values of covariates. Studying the properties of individual curves allows us to determine if the classification methods are good enough for our purposes. That is, it enables us to answer questions such as whether medical diagnosis is sufficiently accurate, whether a database search engine identifies sufficiently many of the relevant documents, whether a spam filter allows through few enough spam messages while rejecting very few genuine ones, and so on. But often we need to go beyond absolute performance, and need to compare methods and their resulting ROC curves. We might need to know which of two methods is better, so that we can decide which method to use, or we might want to compare the results of the same classifier when applied to different populations. To investigate such issues, we can choose an appropriate measure of performance, defined on the basis of the ROC curves, and rank order our methods in terms of this measure. But even this is not enough. We need to supplement the ranking by using appropriate statistical tests, so that we can be confident that any apparent superiority was unlikely to have arisen by chance.

Earlier chapters have reviewed summaries of ROC curves, so this chapter is largely concerned with the statistical tests for different measures. First, however, there is a fundamental issue which we must address. This is that *ROC curves can cross*. Figure 3.2 provided an illustration of this: the ROC curves for a quadratic discriminant classifier and a neural network applied to the same data set are shown on the same graph. For *fp* values between about 0.1 and 0.3, quadratic discriminant analysis produces the larger *tp* values. However, for *fp* outside this range, the neural network produces the larger *tp* values. That is, for *fp* in the approximate range [0.1, 0.3], quadratic discriminant analysis is superior, but otherwise the neural network is better. This problem has already been mentioned in Section 3.5.4. Since different *fp* values correspond to different values of the threshold t, this also tells us that for some threshold values quadratic discrimination is better and for other values the neural network is better.

If we know the value of t which will be used in practice, then there is no difficulty. Equivalently, if we know the value of *fp* (or *tp*) that we wish to use in practice, then there is no difficulty. But for those many problems where t (or other aspects) are not pre-determined, then difficulties arise. For example, the AUC has the advantage that it does not require a value of t to be specified. It aggregates performance over the entire range of t values, summarizing the entire ROC curve. But this very advantage is also its disadvantage when the curves cross. In particular, one can easily conjure up examples in which the AUC for classifier 1 is larger than the AUC for classifier 2, even though classifier 2 is superior to classifier 1 for almost all choices of the classification threshold t. It is precisely this difficulty which is one of the motivations behind the use of the partial area under the curve and the LC statistic described in Section 3.5. These measures suppose that although precise values of t or *fp* are not known, something is known about their likely values.

Bearing such complications in mind, there are various factors we must take into account when comparing ROC curves. These include the following.

Firstly, we have to decide whether we wish to make overall statements about the curves (e.g., that at *no* value of the threshold t are curves significantly different), or about specific summary statistics (e.g., AUC, PAUC).

Secondly, we might be interested in comparing a pair of curves, or we might have more than two to compare. Most of this chapter concentrates on the former case because it is the most important one, but there is also some discussion of more general situations. In particular, the regression modeling and glm modeling approaches permit ready extension to arbitrary numbers of curves, related in various ways.

Thirdly, comparisons will be based on performance of the classifiers applied to data sets. Now, the same data set might be used to produce each of the ROC curves, or different data sets might be used for different curves. That is, the comparison might be *paired* or *unpaired*. If the same data sets are used (paired), then the ROC curves, and any summary statistics produced from them, will be correlated. Failure to take account of the fact that the same data set has been used to construct ROCs when comparing them is a not uncommon mistake. It means that a test for whether curves are different is likely to be less powerful than one which allows for the correlation. Occasionally one also encounters situations in which some of the objects in the sample being studied have been assigned to classes by both classifiers, while others have been assigned by just one of them.

Fourthly, the curves may have been estimated parametrically (e.g., the binormal model discussed in earlier chapters) or nonparametrically.

Fifthly, the sampling procedure will influence the variance and covariances of the curves. For example, sometimes individual objects may yield more than one score, so that the scores are clustered. Since, in general, within-object variability is smaller than between-object variability, it is necessary to allow for this in the analysis.

It will be clear from the above that, no matter how comprehensive we might seek to be in covering situations and complications which might arise, we will not be able to predict all of them. In this chapter we look at some of the most important special cases. In other cases, particular methods may need to be developed. In complicated situations general resampling methods such as the bootstrap approach can be applied.

6.2 Comparing summary statistics of two ROC curves

We introduced summary statistics for individual ROC curves in Chapter 2. Important such statistics include the area under the curve (the AUC), the partial area under the curve (PAUC), the values of the curves at a particular point (e.g., the tp values for a particular value of the fp; the tp values for a fixed value of the threshold t; the misclassification rate for a fixed threshold t), the maximum vertical distance between the ROC curve and the chance diagonal, and the cost-weighted misclassification rate. ROC curves can then be compared by comparing their summary statistics.

Various tests have been developed for specific summary statistics, and we discuss some of the more important ones below. However, a general approach can be based on the regression modeling strategy described in Section 5.3 when adjusting for covariates. If a categorical covariate is used, with different values corresponding to the summary statistic of different ROC curves, then such approaches offer an immediate way to compare summary statistics: one simply tests the relevant regression coefficients against a null hypothesis that they have value 0.

Turning to particular tests, several summary statistics reduce to proportions for each curve, so that standard methods for comparing proportions can be used. For example, the value of tp for fixed fp is a proportion, as is the misclassification rate for fixed threshold t. We can thus compare curves, using these measures, by means of standard tests to compare proportions. For example, in the unpaired case, we can use the test statistic $(tp_1 - tp_2)/s_{12}$, where tp_i is the true positive rate for curve i at the point in question, and s_{12}^2 is the sum of the variances of tp_1 and tp_2. Some care is required, however, to ensure that the correct variances are used. In particular, as we saw in Section 3.4, if the performance criterion is the value of tp for a fixed threshold t, then the variance of tp is given by the standard binomial variance of the proportion, but if the performance criterion is the value of tp for a fixed value of fp, then the variance of tp is inflated by the uncertainty in the estimation of fp. That is, for fixed fp the variance of tp for each of the classifiers is

$$\frac{tp \times (1-tp)}{n_P} + \left(\frac{g(c)}{f(c)}\right)^2 \frac{fp \times (1-fp)}{n_N}$$

where $g(c)$ is the pdf of scores for class P, $f(c)$ is the pdf of scores for class N, c is given by inverting $fp = 1 - F(c)$, and $F(c)$ is the cdf of scores for class N.

For the paired case, one can derive an adjustment to allow for the covariance between tp_1 and tp_2, but an alternative is to use McNemar's test for correlated proportions (Marascuilo and McSweeney, 1977).

For the binormal case, Greenhouse and Mantel (1950) describe a test to compare the sensitivities (true positive rates) of two classifiers for a given common level of specificity (1−false positive rate). For unpaired classifiers, this is as follows. Let $z_{1-\beta}$ be the standard normal deviate corresponding to the level of specificity, β, at which we wish to

6.2 COMPARING SUMMARY STATISTICS

compare the sensitivities of the classifiers. Then

$$T = \sigma_P^{-1} \left(\mu_P - \mu_N - z_{1-\beta}\sigma_N \right)$$

gives the corresponding sensitivity. The classifier which has the greater value of T is the better.

A sample estimate of T is given by

$$\hat{T} = s_P^{-1} \left(\bar{x}_P - \bar{x}_N - z_{1-\beta}s_N \right)$$

where $\bar{x}_P$ is the mean of the class P scores, $\bar{x}_N$ is the mean of the class N scores, s_P is the sample standard deviation of the class P scores, and s_N is the sample standard deviation of the class N scores. The variance of $\hat{T}$ is approximately

$$V\left(\hat{T}\right) \approx \frac{T^2}{2n_P} + \frac{1}{n_P} + \frac{\sigma_N^2}{n_N \sigma_P^2} + \frac{(z_{1-\beta}\sigma_N)^2}{2n_N \sigma_P^2}$$

from which an estimate of the variance of $\hat{T}$ can be obtained on replacing the population quantities by their sample estimates.

Denoting the values of $\hat{T}$ for the two classifiers by $\hat{T}_1$ and $\hat{T}_2$, a test statistic for equality of the sensitivities at the given common specificity value β is then given by

$$\frac{\hat{T}_1 - \hat{T}_2}{\sqrt{\hat{V}\left(\hat{T}_1 - \hat{T}_2\right)}} = \frac{\hat{T}_1 - \hat{T}_2}{\sqrt{\hat{V}\left(\hat{T}_1\right) + \hat{V}\left(\hat{T}_2\right)}}$$

which can be compared with the appropriate critical value of the standard normal distribution.

For paired classifiers, it is necessary to allow for the covariance between the scores obtained by the two classifiers, and Greenhouse and Mantel (1950) suggest replacing $\hat{V}\left(\hat{T}_1 - \hat{T}_2\right)$ in the above by

$$\begin{aligned}
\hat{V}\left(\hat{T}_1 - \hat{T}_2\right) &= \frac{1}{2n_P}\left(\hat{T}_1^2 + \hat{T}_2^2 - 2r_P^2\hat{T}_1\hat{T}_2 - 4r_P + 4\right) \\
&+ \frac{2 + z_{1-Sp}^2}{2n_N}\left(\frac{s_{N1}^2}{s_{P1}^2} - \frac{s_{N2}^2}{s_{P2}^2}\right) \\
&- \frac{s_{N1}s_{N2}}{n_N s_{P1} s_{P2}}\left(2r_N + z_{1-Sp}^2 r_N^2\right)
\end{aligned}$$

where r_P is the sample correlation between scores for classifiers 1 and 2 for the class P cases and r_N is the sample correlation between scores for classifiers 1 and 2 for the class N cases.

Bloch (1997) describes tests for comparing two classifiers on the basis of overall misclassification cost using the same data set. Following Section 2.3.1, the expected cost of misclassification with classification threshold t is

$$C = q \times [1 - tp] \times c(\text{N}|\text{P}) + (1 - q) \times fp \times c(\text{P}|\text{N}) \quad (1)$$

where q is the proportion of objects in class P, and $c(i|j)$ is the cost of misclassifying a class j object as class i, $i \neq j$ (where we have assumed that correct classifications incur no cost). This will lead to an estimate

$$\hat{C} = \hat{q} \times \left[1 - \widehat{tp}\right] \times c(\text{N}|\text{P}) + (1 - \hat{q}) \times \widehat{fp} \times c(\text{P}|\text{N}). \quad (2)$$

A test statistic to compare the costs C_1 and C_2 of classifiers 1 and 2 is then given by

$$Z = \frac{\hat{C}_1 - \hat{C}_2}{\sqrt{\hat{V}\left(\hat{C}_1\right) + \hat{V}\left(\hat{C}_2\right) - 2\widehat{Cov}\left(\hat{C}_1, \hat{C}_2\right)}}.$$

The variances $V\left(\hat{C}_i\right)$ and covariance $Cov\left(\hat{C}_1, \hat{C}_2\right)$, and hence also their estimates, will depend on the sampling procedures adopted in (2). For example, if N objects are randomly drawn from the complete population, then covariances are induced between

$$\hat{q} \times \left[1 - \widehat{tp}\right]$$

and

$$(1 - \hat{q}) \times \widehat{fp}.$$

In this case, Bloch (1997) gives the variances and covariances as

$$nV\left(\hat{C}_i\right) = c(\text{N}|\text{P})^2 p(s_i \leq t|\text{P}) + c(\text{P}|\text{N})^2 p(s_i > t|\text{N}) - \hat{C}_i^2$$

where $p(s_i > t)$ is the probability that the score for the ith classifier i will be larger than t, and

$$\begin{aligned} nCov\left(\hat{C}_1, \hat{C}_2\right) &= c(\text{N}|\text{P})^2 p(s_1 \leq t,\ s_2 \leq t|\text{P}) \\ &+ c(\text{P}|\text{N})^2 p(s_1 > t,\ s_2 > t|\text{N}) - \hat{C}_1 \times \hat{C}_2. \end{aligned}$$

The estimates can be obtained on replacing the probabilities by sample proportions.

6.3 Comparing AUCs for two ROC curves

For the case of uncorrelated ROC curves not based on any parametric distributional assumptions, statistics for tests and confidence intervals comparing AUCs can be straightforwardly based on estimates $\widehat{AUC}_1$ and $\widehat{AUC}_2$ of the AUC values, and estimates of their standard deviations S_1 and S_2, as given in Section 3.5.1:

$$Z = \frac{\widehat{AUC}_1 - \widehat{AUC}_2}{\sqrt{S_1^2 + S_2^2}}$$

One of the earliest proposals for such a test was given by Hanley and McNeil (1982), who used the form of the variance given by Bamber (1975) (see Section 3.5.1).

To extend such tests to the case in which the same data set is used for both classifiers, we need to take account of the correlation between the AUC estimates:

$$z = \frac{\widehat{AUC}_1 - \widehat{AUC}_2}{\sqrt{S_1^2 + S_2^2 - rS_1S_2}}$$

where r is the estimate of this correlation. Hanley and McNeil (1983) made such an extension, basing their analysis on the binormal case, but only gave a table showing how to calculate the estimated correlation coefficient r from the correlation r_P of the two classifiers within class P, and the correlation r_N of the two classifiers within class N, saying that the mathematical derivation was available on request. Various other authors (e.g., Zou, 2001) have also developed tests based on the binormal model, assuming that an appropriate transformation can be found which will simultaneously transform the score distributions of classes P and N to normal.

DeLong *et al.* (1988) took advantage of the identity between AUC and the Mann-Whitney test statistic, together with results from the theory of generalized U-statistics due to Sen (1960), to derive an estimate of the correlation between the AUCs that does not rely on the binormal assumption. In fact, DeLong *et al.* (1988) presented the following results for comparison between $k \geq 2$ classifiers.

In Section 3.5.1, we showed that the area under the empirical ROC curve was equal to the Mann-Whitney U-statistic, and was given by

$$\widehat{AUC} = \frac{1}{n_N n_P} \sum_{i=1}^{n_N} \sum_{i=1}^{n_P} \left[I(s_{Pj} > s_{Ni}) + \frac{1}{2} I(s_{Pj} = s_{Ni}) \right],$$

where s_{Pj}, $j = 1, ..., n_P$ are the scores for the class P objects and s_{Ni}, $i = 1, ..., n_N$ are the scores for the class N objects in the sample. Suppose that we have k classifiers, yielding scores s_{Nj}^r, $j = 1, ..., n_N$, s_{Pi}^r, $i = 1, ..., n_P$, and $\widehat{\text{AUC}}_r$, $r = 1, ..., k$. Define

$$V_{10}^r(s_{Pi}) = \frac{1}{n_N} \sum_{j=1}^{n_N} \left[I\left(s_{Pi}^r > s_{Nj}^r\right) + \frac{1}{2} I\left(s_{Pi}^r = s_{Nj}^r\right) \right], \quad i = 1, ..., n_P$$

and

$$V_{01}^r(s_{Nj}) = \frac{1}{n_P} \sum_{i=1}^{n_P} \left[I\left(s_{Pi}^r > s_{Nj}^r\right) + \frac{1}{2} I\left(s_{Pi}^r = s_{Nj}^r\right) \right], \quad j = 1, ..., n_N.$$

Next define the $k \times k$ matrix $\mathbf{W}_{10}$ with (r,s)th element

$$w_{10}^{r,s} = \frac{1}{n_P - 1} \sum_{i=1}^{n_P} \left[V_{10}^r(s_{Pi}) - \widehat{\text{AUC}}_r \right] \left[V_{10}^s(s_{Pi}) - \widehat{\text{AUC}}_s \right]$$

and the $k \times k$ matrix $\mathbf{W}_{01}$ with (r,s)th element

$$w_{01}^{r,s} = \frac{1}{n_N - 1} \sum_{i=1}^{n_N} \left[V_{01}^r(s_{Ni}) - \widehat{\text{AUC}}_r \right] \left[V_{01}^s(s_{Ni}) - \widehat{\text{AUC}}_s \right].$$

Then the estimated covariance matrix for the vector $\left(\widehat{\text{AUC}}_1, ..., \widehat{\text{AUC}}_k\right)$ of estimated areas under the curves is

$$\mathbf{W} = \frac{1}{n_P} \mathbf{W}_{10} + \frac{1}{n_N} \mathbf{W}_{01}$$

with elements $w^{r,s}$. This is a generalization of the result for the estimated variance of a single estimated AUC, also given in Section 3.5.1. In the case of two classifiers, the estimated correlation r between the estimated AUCs is thus given by $w^{1,2} / \sqrt{w^{1,1} w^{2,2}}$, which can be used in z above.

As we showed in Section 2.4.1, the area under the curve is the average sensitivity, assuming that each value of the specificity is equally likely. This leads to an immediate generalization, in which one assumes a weight distribution for the specificity values. The partial area under the curve is, of course, an example of this generalization. Wieand et al. (1989) give an asymptotic derivation of the variance of the difference between two such weighted areas, for arbitrary weight function, with the variance of the standard AUC and the variance of the sensitivity at a particular specificity value being special cases. They also give results for the standard AUC in the binormal case.

6.4 Comparing entire curves

So far we have discussed comparisons of summary statistics of ROC curves. Summary statistics are all very well—they reduce things to a single numerical scale so that comparisons can be made. However, the very reduction of a complex shape to a single numerical value inevitably means that information is being sacrificed. Whether this is appropriate or not depends, of course, on one's aim, but sometimes it will certainly be inappropriate. In such cases, we need to compare entire curves. Broadly speaking, we can separate these approaches into two kinds, those based on regression modeling of the entire curve, and those based on nonparametric approaches. A particular special case of the former arises in the context of the familiar binormal model; so, to introduce the ideas, we discuss this case first. We then examine nonparametric approaches, before returning to the more general regression approach.

6.4.1 The binormal case

Let the mean and standard deviation of the score distribution for ROC curve i, $i = 1, 2$, in population P be represented by μ_{Pi} and σ_{Pi} respectively, with analogous representations for the mean of population N. As we showed in Section 2.5, a convenient reparameterization of the ROC curve is given by $a_i = (\mu_{Pi} - \mu_{Ni})/\sigma_{Pi}$ and $b_i = \sigma_{Ni}/\sigma_{Pi}$, since then the ROC curve takes the form $y = \Phi\left(a + b\Phi^{-1}(x)\right)$. This reduces the previous eight parameters (the mean and standard deviation for each of populations P and N in each of ROC curves 1 and 2) to just four. Denote the difference between the a_i parameters by $a_{12} = a_1 - a_2$ and the difference between the b_i parameters by $b_{12} = b_1 - b_2$. Then our aim is to test the null hypothesis that simultaneously $a_{12} = 0$ and $b_{12} = 0$ against the alternative hypothesis that $a_{12} \neq 0$ and/or $b_{12} \neq 0$. A suitable test statistic is given by

$$X^2 = \frac{\hat{a}_{12} V\left(\hat{b}_{12}\right) + \hat{b}_{12}^2 V\left(\hat{a}_{12}\right) - 2\hat{a}_{12}\hat{b}_{12} C\left(\hat{a}_{12}, \hat{b}_{12}\right)}{V\left(\hat{a}_{12}\right) V\left(\hat{b}_{12}\right) - C\left(\hat{a}_{12}, \hat{b}_{12}\right)^2} \quad (3)$$

in which $\hat{a}_{12}$ is the sample estimate of a_{12}, $\hat{b}_{12}$ is the sample estimate of b_{12},

$$V(\hat{a}_{12}) = \hat{\sigma}_{a_1}^2 + \hat{\sigma}_{a_2}^2 - 2\hat{\sigma}_{a_1 a_2}, \quad (4a)$$

$$V\left(\hat{b}_{12}\right) = \hat{\sigma}_{b_1}^2 + \hat{\sigma}_{b_2}^2 - 2\hat{\sigma}_{b_1 b_2}, \quad (4b)$$

and
$$C\left(\hat{a}_{12}, \hat{b}_{12}\right) = \hat{\sigma}_{a_1 b_1} + \hat{\sigma}_{a_2 b_2} - \hat{\sigma}_{a_1 b_2} - \hat{\sigma}_{a_2 b_1}, \quad (4c)$$

where $\hat{\sigma}_X$ denotes the sample standard deviation of X, $\hat{\sigma}_X^2$ its sample variance, and $\hat{\sigma}_{XY}$ the sample covariance between X and Y. The test statistic in (3) asymptotically follows a chi-squared distribution with two degrees of freedom if the null hypothesis is true.

In the case in which independent samples are used to construct the two ROC curves, expressions (4a), (4b), and (4c) simplify, with the covariance terms between the different curves, $\hat{\sigma}_{a_1 a_2}$, $\hat{\sigma}_{b_1 b_2}$, $\hat{\sigma}_{a_1 b_2}$, and $\hat{\sigma}_{a_2 b_1}$ becoming zero.

6.4.2 Nonparametric approach

In Section 6.4.1 we assumed that the two ROC curves to be compared each had a normal form. While this is sometimes a justifiable assumption (perhaps after transformation) it is often not (perhaps the distributions cannot be simultaneously transformed to normality). We therefore require more general methods. More general tests of equality of ROC curves can be based on the following principle. Since ROC curves are invariant to monotonic transformations of the classification threshold, we will obtain the same curve after applying an arbitrary such transformation. In particular, in principle we could transform the score scale for classifier 2 so that the distribution of class P scores for this classifier was the same as the distribution of class P scores for classifier 1. The ROC curves for the two classifiers would then be identical if, after this transformation of the classifier 2 scale, the distributions of the class N scores were identical for the two classifiers. A nonparametric test of the identity of the two class N distributions could then be applied to test the hypothesis that the two ROC curves were identical.

Venkatraman (2000) has applied this idea, but instead of transforming the class 2 distribution, he implicitly transforms the entire mixture distribution. The argument is as follows. Once again, $f_i(x)$ and $g_i(x)$ are the pdfs of class N and P scores respectively, for classifier i, $i = 1, 2$, with corresponding cdfs $F_i(x)$ and $G_i(x)$. Let $x_{1\pi}$ be the πth quantile for classifier 1, and $x_{2\pi}$ be the πth quantile for classifier 2, so that

$$p(\text{N}) F_1(x_{1\pi}) + (1 - p(\text{N})) G_1(x_{1\pi}) = \pi \quad (5)$$

and

$$p(\text{N}) F_2(x_{2\pi}) + (1 - p(\text{N})) G_2(x_{2\pi}) = \pi. \quad (6)$$

6.4. COMPARING ENTIRE CURVES

Now, when the threshold t is such that a proportion π of the overall population are assigned to class N, the total misclassification error rate for classifier 1 is

$$e_1(\pi) = p(N) F_1(x_{1\pi}) + (1 - p(N))(1 - G_1(x_{1\pi})), \quad (7)$$

and when the threshold t is such that a proportion π of the overall population are assigned to class N, the total misclassification error rate for classifier 2 is

$$e_2(\pi) = p(N) F_2(x_{2\pi}) + (1 - p(N))(1 - G_2(x_{2\pi})). \quad (8)$$

Substituting $F_1(x_{1\pi})$ from (5) into (7) yields

$$e_1(\pi) = \pi - (1 - p(N))(1 - 2G_1(x_{1\pi})), \quad (9)$$

and substituting $F_2(x_{2\pi})$ from (6) into (8) yields

$$e_2(\pi) = \pi - (1 - p(N))(1 - 2G_2(x_{2\pi})), \quad (10)$$

so that

$$e_1(\pi) - e_2(\pi) = 2(1 - p(N))(G_2(x_{2\pi}) - G_1(x_{1\pi})). \quad (11)$$

Thus the misclassification error rate of the two classifiers is equal if and only if the sensitivities of the two classes are the same.

A similar exercise shows that

$$e_1(\pi) = 1 - p(N) - \pi + 2p(N) F_1(x_{1\pi})$$

and

$$e_2(\pi) = 1 - p(N) - \pi + 2p(N) F_2(x_{2\pi})$$

so that

$$e_1(\pi) - e_2(\pi) = 2p(N)(F_1(x_{1\pi}) - F_2(x_{2\pi})), \quad (12)$$

so that the misclassification error rate is the same if and only if the specificities of the two classifiers are the same.

Putting these two results together, the ROC curves are identical if and only if the misclassification rates of the classifiers are the same for all π. This means that we can test the identity of the ROC curves by testing that the integrated unsigned difference between the misclassification rates is zero. That is, by testing

$$\int |e_1(\pi) - e_2(\pi)| \, d\pi = 0. \quad (13)$$

Beginning with unpaired data, let $\hat{F}_1$ and $\hat{G}_1$ denote the sample cumulative distribution functions for classifier 1. We can define

$$\hat{\pi}_i = p(N) \hat{F}_1(x_i) + (1 - p(N)) \hat{G}_1(x_i)$$

and

$$\hat{e}_1(\hat{\pi}_i) = p(N) \hat{F}_1(x_i) + (1 - p(N)) \left(1 - \hat{G}_1(x_i)\right).$$

Defining $\hat{\pi}_0 = 0$ and $\hat{e}_1(\hat{\pi}_0) = 1 - p(N)$, and noting that $\hat{\pi}_n = 1$ and $\hat{e}_1(\hat{\pi}_n) = p(N)$, where N is the number of objects used in constructing the classifier 1 ROC curve, Venkatraman estimates the function $e_1(\pi)$ by joining the points $(\hat{\pi}_i, \hat{e}_1(\hat{\pi}_i))$, $i = 0, ..., n$ using straight lines. A similar approach yields the estimate of the function $e_2(\pi)$ for classifier 2. The integral in (13) is then estimated as

$$\int |\hat{e}_1(\pi) - \hat{e}_2(\pi)| \, d\pi.$$

Venkatraman also describes a procedure for generating a reference distribution for the null hypothesis of no difference between the curves, with which this test statistic can be compared.

Venkatraman and Begg (1996) have developed a corresponding approach for the case of paired data. As we shall see, the pairing makes the test statistic rather more straightforward to calculate.

Suppose that there are N paired scores. Rank the scores for each classifier separately, and suppose that the score on classifier i for the jth object has rank r_{ij}. Let us take a particular rank, k say, as the classification threshold. Then classifier 1 assigns any object which has rank less than or equal to k to class N and any object which has rank above k to class P. Likewise, classifier 2 assigns any object which has rank less than or equal to k to class N and any object which has rank above k to class P.

Since the objects are paired, we have four possible pairwise classifications of the objects from class P:

(a) some of the class P objects which are ranked above k by classifier 1 will also be ranked above k by classifier 2, i.e., $r_{1j} > k$ and $r_{2j} > k$;

6.4. COMPARING ENTIRE CURVES

(b) some will be ranked above k by classifier 1 but not by classifier 2, i.e., $r_{1j} > k$ and $r_{2j} \leq k$;

(c) some will not be ranked above k by classifier 1 but will be by classifier 2, i.e., $r_{1j} \leq k$ and $r_{2j} > k$,

(d) some will not be ranked above k by either classifier, i.e., $r_{1j} \leq k$ and $r_{2j} \leq k$.

From this, we see that the sensitivity of classifier 1 is simply the proportion of class P objects that fall into categories (a) plus (b), and the sensitivity of classifier 2 is the proportion of class P objects that fall into categories (a) plus (c). The sensitivities of the two classifiers will be equal if these two proportions are equal. Since category (a) is common to the two proportions, the sensitivities will be equal if the numbers in (b) and (c) are equal.

An analogous argument applies for specificity, in terms of class N.

Venkatraman and Begg (1996) put these results together to define an error matrix as follows:

For class N points

$$e_{jk} = \begin{cases} 1 & \text{if } r_{1j} \leq k \text{ and } r_{2j} > k \quad (c) \\ -1 & \text{if } r_{1j} > k \text{ and } r_{2j} \leq k \quad (b) \\ 0 & \text{otherwise} \end{cases}$$

For class P points

$$e_{jk} = \begin{cases} 1 & \text{if } r_{1j} > k \text{ and } r_{2j} \leq k \quad (b) \\ -1 & \text{if } r_{1j} \leq k \text{ and } r_{2j} > k \quad (c) \\ 0 & \text{otherwise} \end{cases}$$

Then the sum $e_{.k} = \sum_{j=1}^{n} e_{jk}$ measures similarity between the two ROC curves at the kth rank. It is simply the difference in the total number of errors made by each classifier when rank k is used as the classification threshold. Summing the absolute values of these differences over all k yields an overall measure of difference between the ROC curves:

$$D = \sum_{k=1}^{n-1} |e_{.k}|.$$

Venkatraman and Begg (1996) described two permutation procedures to generate the null distribution for this statistic. The first applies when the scores of the two tests are on the same metric, and the second when they are not. For the first, randomly selected objects have their scores on classifiers 1 and 2 exchanged. The resulting new set of data is then ranked and D is computed. This process is repeated to give the permutation distribution. For the second, randomly selected objects have their ranks on classifiers 1 and 2 exchanged, followed by a random tie breaking step. Again, the results are reranked and the new D calculated, to provide a point from the permutation distribution.

Venkatraman and Begg (1996) compared the statistical properties of this test with those of the test of equality of the area under the curve for correlated data described by DeLong *et al.* (1988) and concluded that, in their simulations, their permutation test had power essentially equivalent to the test of DeLong *et al.* whenever the latter test was appropriate (i.e., when one ROC curve uniformly dominates the other) and was clearly superior in power when neither ROC curve uniformly dominated.

6.4.3 Regression approaches

We introduced regression approaches in Section 5.2, when we discussed methods for modeling the effect of covariates on ROC curves. In that discussion, we noted that there were two distinct approaches: modeling the effect of covariates on the score distributions of the two classes separately, and then combining the results into a model for the ROC curves; or modeling the effect of the covariates on the ROC curve directly. The second of these approaches seems more natural, since the ROC curve describes the relationship between the two score distributions, rather than the distributions themselves—it is, after all, invariant to monotonic transformations of the score scale. For this reason, we restrict the present discussion to this second approach.

In Section 5.2.2, we described the generalized linear model approach to fitting ROC curves, the ROC-GLM model. This models the true positive rate y in terms of the false positive rate x by a generalized linear model

$$h(y) = b(x) + \beta^T \mathbf{Z},$$

where $h(.)$ is the link function, $b(.)$ is a baseline model, both being monotonic on $(0, 1)$, and $\mathbf{Z}$ is a vector of covariates. It is immediately obvious that this model can be used to compare ROC curves. To do this, we index the curves by a categorical covariate Z. For example, in comparing just two ROC curves, Z will be binary, taking the value 0 for one curve and 1 for the other. Then, to test whether the two ROC curves are different, we compare a model which includes the covariate Z with a model without the covariate. If the former provides significantly better fit, then the covariate is necessary—that is, the two curves are different. Of course, in practice we simply fit the model including Z and test to see if its coefficient β can be treated as zero. We have already seen an example of this formulation in Section 5.2.3. This general approach permits immediate generalization to covariates with more than two categories, and to comparisons of ROC curves defined by combinations of covariate values.

6.5 Identifying where ROC curves differ

Summary statistics can be used to tell us whether ROC curves differ in certain ways, and overall tests to compare curves can tell us whether they differ at all. If any such test is positive, however, then it will usually be of interest to determine just where the curves differ. As above, there are various ways in which one can tackle this, but one common way is to find the range of false positive values (the horizontal axis of the ROC curves) on which the true positive values (the vertical axis) of the two curves have significant difference. This range can be found by constructing a simultaneous confidence interval for the difference between the true positive proportions at all values of the false positive proportion. Then those values of false positive proportion for which the confidence interval excludes zero indicate values for which the true positive values are significantly different. For the binormal model, such simultaneous confidence intervals are obtained as follows.

For classifier 1, from Section 2.5, the ROC curve can be expressed as $\Phi^{-1}(y) = a_1 + b_1 \Phi^{-1}(x)$ or $z_y = a_1 + b_1 z_x$. Similarly, for classifier 2, the ROC curve can be expressed as $z_y = a_2 + b_2 z_x$. We are interested in the difference between the z_y values at each value of z_x: $D(z_y) = a_{12} + b_{12} z_x$, defining $a_{12} = a_1 - a_2$ and $b_{12} = b_1 - b_2$ as above. A simultaneous confidence interval for $D(z_y)$ is then given by

$$\hat{D}(z_y) \pm \left(2F_{2,n-4,\alpha} V\left(\hat{D}(z_y)\right)\right)^{1/2} \quad (14)$$

where $F_{2,n-4,\alpha}$ is the $\alpha\%$ upper $\alpha\%$ value for an F distribution with 2 and $n-4$ degrees of freedom, with N the total sample size. This interval can be expressed in terms of the variances and covariances of the estimates of $a_{12} = a_1 - a_2$ and $b_{12} = b_1 - b_2$, defined in Section

6.3.1, so that the confidence interval excludes zero whenever

$$\begin{aligned} W &= z_y^2 \left(\hat{b}_{12}^2 - 2F_{2,n-4,\alpha} V\left(\hat{b}_{12}\right) \right) + 2z_y \left(2F_{2,n-4,\alpha} C\left(\hat{a}_{12}, \hat{b}_{12}\right) - \hat{a}_{12}\hat{b}_{12} \right) \\ &\quad + \left(\hat{a}_{12}^2 - 2F_{2,n-4,\alpha} V\left(\hat{a}_{12}\right) \right) \\ &= Az_y^2 + Bz_y + C \end{aligned}$$

is greater than zero. Since this is a quadratic expression in z_y, we can easily solve $W = 0$ to find the values of z_y for which W is zero, and hence deduce those values for which $W > 0$.

6.6 Further reading

The test of equality of two ROC curves under the binormal assumption was developed by Metz and Kronman (1980) and Metz *et al.* (1984). The nonparametric test for paired curves was described by Venkatraman and Begg (1996). McClish (1990) developed the test for ranges of false positive values for which the true positive values were significantly different.

Sometimes the data are only partially paired, with some objects being scored by two tests and the remainder by just one of the tests. Metz *et al.* (1998) developed a test for comparing the AUCs produced in such cases, based on the binormal model.

Chapter 7

Bayesian methods

7.1 Introduction

Most of the methodology described in the preceding chapters has been developed within the classical ("frequentist") framework of statistical inference. In this framework, populations are represented by probability models whose parameters are treated as fixed but unknown quantities about which inferences are to be made, such inferences are based exclusively on sample data, and the assessment of quantities such as bias, significance or confidence depends on the frequencies with which given sample values might arise if repeated samples were to be taken from the relevant populations. Thus, within this framework there is no scope for incorporating any extraneous information that the researcher might have gleaned about some of the population parameters (say from a series of previous experiments with similar materials), or for incorporating any subjective elements into the analysis.

However, other approaches can be adopted for statistical inference and the most popular one is the Bayesian approach. Here population parameters are treated as random variables possessing probability distributions that reflect the "degree of belief" which the analyst or researcher has about their values, and this therefore permits the introduction of the researcher's prior knowledge or subjective views about the parameters into the analysis. These "prior distributions" are combined with the sample data to produce modified, "posterior" distributions of the parameters, and all inferences about the parameters are then based on these posterior distributions. Thus, for example, a point estimate of a parameter is most naturally given by the mode of its posterior distribution (the "most likely" value), the equivalent of a confidence interval is given by the interval within which the posterior distribution has specified probability content, and a hypothesis test that specifies a range of values for a parameter is conducted by computing the posterior probability of its belonging to this range and assessing whether this probability is sufficiently large for the hypothesis to be tenable.

Such an approach to inference has long been available, but was hampered for many years by the intractability of some of the mathematics and associated computations, specifically the multivariable integration, that the methodology entailed. Since the mid 1990s, however, rapid computational advances using Markov-chain Monte Carlo methods have obviated the need for such multidimensional integration. The idea is basically a simple one. In most cases, the integral to be evaluated can be viewed as the expectation of some function $f(\cdot)$ over the posterior distribution of the population parameters, so if a sample of values can be generated from this posterior distribution then the corresponding sample values of $f(\cdot)$ can be obtained and the integral can be approximated by the average of the latter values. Traditional methods of sampling (uniform, rejection, or importance) cannot usually cope with the problems presented in most Bayesian situations, but the Markov-chain variants of importance sampling ensure efficient and principled sampling from the appropriate distribution. The Markov chain feature is that each new proposal depends on the current one, and the sequence of proposals is guaranteed to converge to values from the desired distribution. The two popular algorithms are Metropolis-Hastings and Gibbs sampling; the former uses the joint distribution of new and current proposals, while the latter uses a sequence of conditional distributions. All-purpose software such as WinBUGS (Spiegelhalter *et al.*, 1995) is now widely available and has made Bayesian methods very popular.

ROC analysis is one of the application areas to have benefitted from these advances, but the developments have tended to focus on just a few specific aspects. To date there are still relatively few Bayesian methods to complement the classical approaches in most of the general areas described in foregoing chapters, although research in these areas is beginning to increase. However, Bayesian methodology does come into its own in situations where there is some uncertainty about underlying quantities, and ROC analysis throws up several such situations. One arises when it is desired to pool the information from several related studies but there are varying degrees of confidence about the accuracy or reliability of the results in each one. This general situation is termed "meta-analysis," and as well as some classical approaches there is now a well-established Bayesian route to the analysis. Another problem area comes when the labeling as N or P of subjects from which a ROC curve is to be constructed is either fallible or not available. This can happen in various fields but arguably most often in medical studies, where the ROC curve is required for a disease diagnostic test but the true disease status of each sample member is either equivocal or unknown and ways of allowing for this uncertainty have to be introduced. Again, Bayesian methods have proved to be very useful in developing appropriate analyses.

In this chapter we consider each of these areas, and survey the various methods that have been developed to deal with them.

7.2 General ROC analysis

Bayesian modeling has been applied successfully to many classification or diagnostic testing problems in standard situations when the classifier score S is either binary (e.g., Dendukuri and Joseph, 2001) or ordinal (Peng and Hall, 1996; Hellmich et al., 1998), but there is little material available to guide the Bayesian enthusiast when S is continuous. Indeed, the only specific proposal appears to be the one by Erkanli et al. (2006) for Bayesian semi-parametric ROC analysis, so we describe their approach here.

They start by noting that in areas such as medical testing, the classification scores rarely have normal, or even symmetric, distributions. Hence the binormal model (Section 2.5), where the scores S are assumed to be normally distributed with means μ_P, μ_N ($\mu_P > \mu_N$) and standard deviations σ_P, σ_N in populations P and N respectively, is not a good basis for analysis and some more general form is needed. In a Bayesian context, mixtures of normals have proved to be very flexible in a number of modeling tasks (Ferguson, 1983; Escobar and West, 1995; Müller et al., 1996), so Erkanli et al. (2006) build on this previous work. To avoid overly complex suffix notation, let us for the moment ignore the population markers N and P, and consider the distribution of the classification score S in either of these populations.

Suppose that C denotes the number of components in the normal mixture, and let K be the random variable indicating the operative component for a given classification score. Then for a Bayesian approach we write the model formally as

$$S|K, \theta_K \sim N(\mu_K, \sigma_K^2)$$

where K and $\theta_K = (\mu_K, \sigma_K^2)$ are parameters so must be assigned prior distributions. Erkanli *et al.* (2006) provide a perfectly general framework for these distributions, but for computational ease work in terms of some standard conjugate-family specifications. In particular, the components of θ_K are assigned conjugate normal-gamma baseline priors (the means having normal distributions independently of the inverse variances which have gamma distributions) while the component index K has an independent C-state multinomial distribution whose probabilities $w_1, \ldots, w_C$ are specified as

$$w_1 = R_1, \quad w_k = (1 - R_1)(1 - R_2) \ldots (1 - R_{k-1}) R_k \quad \text{for} \quad k = 2, \ldots, C$$

where the R_i are independent Beta$(1, \alpha)$ variables and $R_C = 1$ to ensure that the w_i sum to 1. The value of the parameter α has implications for the number of components C, and Erkanli *et al.* (2006) discuss possible choices. Ishwaran and James (2002) identify this model as one that arises from a finite approximation to a Dirichlet process, so the prior distributions are termed *mixed Dirichlet process* (MDP) priors.

Suppose now that a set of scores $s_1, s_2, \ldots, s_n$ has been observed, and we are interested in predicting the value s of a future score. According to the above model, the density function of s is a finite mixture of the form

$$f(s|w_i, \theta_i; i = 1, \ldots, C) = \sum_{i=1}^{C} w_i f(s|\theta_i),$$

where the $f(s|\theta_i)$ for $i = 1, \ldots, C$ are normal densities. Thus if $g(\boldsymbol{w}, \boldsymbol{\theta}|s) = g(w_i, \theta_i; i = 1, \ldots, C | s_1, \ldots, s_n)$ is the posterior density of the w_i, θ_i, then the posterior predictive density of the future score s given the past scores s_i $(i = 1, \ldots, n)$ is

$$\begin{aligned} f(s|s_1, \ldots, s_n) &= \int \cdots \int f(s|w_i, \theta_i; i = 1, \ldots, C) \times \\ &\quad g(\boldsymbol{w}, \boldsymbol{\theta}|s) dw_1 \cdots dw_C d\theta_1 \cdots d\theta_C, \end{aligned}$$

which is just the expectation of $f(s|w_i, \theta_i; i = 1, \ldots, C)$ taken over the posterior distribution of the w_i, θ_i. Exact computation of this integral

7.2. GENERAL ROC ANALYSIS

or expectation is difficult even for small C, but a Gibbs sampler can be developed in order to simulate U observations $(w_i^u, \theta_i^{(u)}; i = 1, \ldots, C)$ for $u = 1, \ldots, U$ from the posterior distribution, and the expectation can then be approximated by the average

$$\frac{1}{U} \sum_{u=1}^{U} \left[\sum_{i=1}^{C} w_i^{(u)} f(s|\theta_i^{(u)}) \right]$$

of these Monte Carlo samples. Erkanli et al. (2006) give closed analytical expressions for all the conditional distributions that are needed for the Gibbs sampler, and hence provide the requisite sampling algorithm. Having approximated the posterior predictive density $f(s|s_1, \ldots, s_n)$, it is then a straightforward matter to obtain the cumulative distribution function $F(t|s_1, \ldots, s_n) = \int_{-\infty}^{t} f(s|s_1, \ldots, s_n) ds$.

The above procedure can thus be conducted separately on a sample of n scores $s_1^{(N)}, \ldots, s_n^{(N)}$ from population N to give $F_N(t|s_1^{(N)}, \ldots, s_n^{(N)})$, and then on m scores $s_1^{(P)}, \ldots, s_m^{(P)}$ to give $F_P(t|s_1^{(P)}, \ldots, s_m^{(P)})$ where t is the classifier threshold value. The (posterior predicted) true positve rate is thus

$$tp(t) = 1 - F_P(t|s_1^{(P)}, \ldots, s_m^{(P)}),$$

and the (posterior predicted) false positive rate is

$$fp(t) = 1 - F_N(t|s_1^{(P)}, \ldots, s_m^{(P)}).$$

Varying the threshold t and plotting $tp(t)$ against $fp(t)$ yields the predictive ROC curve, and the area under it can be readily derived. Erkanli et al. (2006) obtain finite mixture form expressions for $tp(t)$, $fp(t)$, and AUC, which facilitates their computation under the Markov chain Monte Carlo process. Moreover, posterior uncertainties relating to all these quantities can be assessed directly from the simulations generated by the Gibbs sampling procedure by means of prediction intervals. Erkanli et al. (2006) illustrate this Bayesian approach, investigate the effects of different choices of parameters and prior distributions, and demonstrate its accuracy and efficiency on several real and simulated data sets.

7.3 Meta-analysis

7.3.1 Introduction

Much scientific research proceeds by a series of small steps, and hence very frequently the need arises for synthesizing results from a set of related but statistically independent research studies. In recent years this need has manifested itself most strongly in medical and pharmaceutical research, where breadth of applicablity of results across populations is of prime importance, but similar requirements arise in many other fields as well. Within the area of classification (or equivalently diagnostic testing), a systematic review and synthesis of all published information about a classifier is necessary for an overall assessment of its value.

When considering such published information, however, the decision maker must only include for synthesis those studies that in general terms can be deemed to be "comparable," and to reject those that either bear on different questions or that have very different background conditions from the rest. Once a suitable set of studies has been identified, traditional methodology requires them first to be categorized as being within either a "fixed-effects" or a "random-effects" framework. Hellmich *et al.* (1999) distinguish between the two frameworks by defining the fixed-effects framework as one in which the studies are treated as identical repeats of each other, while the random-effects framework is one in which the studies deal with the same general question but with some differences from study to study. This is in keeping with the traditional uses of these two terms in the analysis of designed experiments: in both frameworks there is random variation among the individuals in each study, but random effects permit additional variation between studies by treating them as "exchangeable," in the sense that they have been drawn randomly from one overarching population, while fixed effects assume that every study is essentially the same and hence there is no additional between-study variation.

For reviews of meta-analysis in general, and important meta-analytic methods within the classification area, see Irwig *et al.* (1995), Hellmich *et al.* (1999), and Normand (1999); we draw in particular on the latter two in the sections below.

7.3.2 Frequentist methods

The basic premiss is that we wish to summarize the findings from a set of M independent studies with related objectives, each of which has reported some response y. For the continuous classifier situations that we are concerned with, the most likely such responses would be AUC, PAUC, or perhaps one of the other summary ROC indices mentioned in Section 2.4.3. The basic model for the fixed effects meta-analysis assumes that the individual responses y_i are normally distributed with unknown means μ_i but known variances σ_i^2 for $i = 1, 2, \ldots, M$. Hellmich et al. (1999) note that the assumption of known "within-study" variances is widely made, and comment that it is often a good approximation for practical purposes.

For this model, and under the homogeneity hypothesis $H_0 : \mu_1 = \mu_2 = \ldots = \mu_M = \mu$, say, the common response μ is simply estimated by

$$\hat{\mu} = \left(\sum_{i=1}^{M} w_i y_i\right) / \left(\sum_{i=1}^{M} w_i\right)$$

with $\text{Var}(\hat{\mu}) = 1/\left(\sum_{i=1}^{M} w_i\right)$, where $w_i = 1/\sigma_i^2$. The homogeneity assumption may be tested by using Cochran's statistic $Q = \sum_{i=1}^{M} w_i (y_i - \hat{\mu})^2$, which follows a chi-squared distribution on $M - 1$ degrees of freedom if H_0 is true.

If the assumption of homogeneity is not considered to be reasonable, or if the restrictions of the fixed-effects analysis are deemed to be too great, then the random-effects model is more appropriate. Here we again assume $y_i \sim N(\mu_i, \sigma_i^2)$, but now we further assume that $\mu_i \sim N(\mu, \sigma^2)$ where μ and σ^2 are unknown "hyperparameters." Thus the between-study variability is explained by assuming that the observed studies comprise a sample of possible studies that might be conducted. The random-effects analogue of the above fixed-effects analysis requires the calculation of $\hat{\mu}$ and $\text{Var}(\hat{\mu})$ for this model.

There are various routes whereby frequentist estimates can be obtained, the prevalent ones in practice being the methods of maximum likelihood (ML), restricted maximum likelihood (REML), or moments (MM). For either ML or REML, the likelihood of the data is given by

$$L(\mu, \sigma^2) = (2\pi)^{-M/2} \left(\prod_{i=1}^{M} \frac{1}{\sqrt{\sigma_i^2 + \sigma^2}}\right) \exp\left(-\frac{1}{2} \sum_{i=1}^{M} \frac{(y_i - \mu)^2}{\sigma_i^2 + \sigma^2}\right).$$

Taking logs, differentiating the resulting expression successively with respect to μ and σ, and setting each derivative to zero yields a pair of simultaneous equations for the estimates $\hat{\mu}$ and $\hat{\sigma}^2$. The ML estimate is obtained by direct iteration over these equations, while for the REML estimate one of the equations has to be slightly modified before the iteration. Full details are given in appendix A of Hellmich et al. (1999), together with expressions for the estimated variance of $\hat{\mu}$. The method of moments, by contrast, yields a much simpler and non-iterative estimator for σ^2, given by DerSimonian and Laird (1986) as the larger of 0 and $(Q - M + 1) \div (\sum_i w_i - [\sum_i w_i^2]/[\sum_i w_i])$ where w_i and Q are as defined above. Substitution of this estimate for σ^2 into the ML equation for $\hat{\mu}$ then yields a noniterative estimate of μ. Since this is a very simple method, it has been widely applied in practice. Improvements to these basic procedures have been provided by Hardy and Thompson (1996), who show how confidence intervals can be obtained for both parameters, and by Biggerstaff and Tweedie (1997), who refine the distribution of Q and hence produce a new estimate for μ as well as confidence intervals for σ^2.

7.3.3 Bayesian methods

Instead of the frequentist approach of estimating the hyperparameters μ and σ^2 of the random-effects model from the data alone, a Bayesian approach treats the parameters as variables, posits prior distributions for them, and conducts inference on the basis of these distributions. However, there are several ways of proceeding: one which steers a course mid-way between frequentist and Bayesian approaches, known as the Empirical Bayes approach, and one which fully embraces the Bayesian philosophy.

Empirical Bayes

Here the usual procedure is to posit a noninformative uniform conditional prior $f(\mu|\sigma)$ for μ, assuming σ to be fixed and given. Standard Bayesian theory then establishes that the resulting posterior conditional distributions of μ and the μ_i for $i = 1, 2, \ldots, M$ are all normal. Moreover, if we write $u_i = 1/(\sigma_i^2 + \sigma^2)$, $v_i = u_i \sigma^2$, $S_u = \sum_i u_i$, and $S_{uy} = \sum_i u_i y_i$, then these distributions have the following moments:

$$E(\mu|\sigma^2, y_1, \ldots, y_M) = S_{uy}/S_u,$$

$$\text{Var}(\mu|\sigma^2, y_1, \ldots, y_M) = 1/S_u;$$
$$E(\mu_i|\sigma^2, y_1, \ldots, y_M) = v_i y_i + (1 - v_i) S_{uy}/S_u,$$
$$\text{Var}(\mu|\sigma^2, y_1, \ldots, y_M) = v_i \sigma^2 + (1 - v_i)^2/S_u.$$

The Empirical Bayes approach is to estimate σ^2 from the $y_1, \ldots, y_M$ and treat the resulting estimate as the "known" value of σ^2. Inserting this value into the above equations enables all the moments to be calculated, and the results are then treated in a frequentist manner as in the previous sections. Further simplification can be achieved by conditioning on both μ and σ^2 when finding the posterior moments of the μ_i, and estimating μ as well as σ^2 from the data, but this is not usually done.

Full Bayes

One drawback with the Empirical Bayes approach is the same as that encountered in many of the frequentist approaches: using the hyperparameter estimate(s) as if they were the true value(s) ignores the uncertainty associated with these estimates. In the fully Bayesian approach a joint prior distribution $f(\mu, \sigma^2)$ is first specified for both hyperparameters, perhaps most naturally by forming the product of the conditional prior $f(\mu|\sigma^2)$ for the mean and the marginal prior $f(\sigma^2)$ for the variance. This joint prior distribution is then combined with the sampling distribution of the data $y_1, \ldots, y_M$ (as specified by the random-effects model) to give the joint posterior distribution of all unknown parameters, viz. $\mu_1, \ldots, \mu_M, \mu$, and σ^2. The posterior predictive density for a new (unobserved) response y is then given by integrating the product of the joint posterior distribution and the normal density $N(\mu, \sigma^2)$ over all the unknown parameters.

Stated baldly as above it all seems to be straightforward, but of course there are many difficulties. One is the need to specify appropriate prior distributions in order to accurately represent the prior knowledge (or lack of it) about the parameters. There are many possible distributions from which the choice can be made, and considerable experience is usually required for good selection. Some sensitivity analysis should always also be conducted, to establish whether the results of the analysis are unduly affected by slight variations in the choice of priors. Another main historical source of problems is in the computing necessary to conduct the multidimensional integrals. Fortunately, these difficulties have been overcome to a great extent with the current easy availability of Markov-chain Monte Carlo software, but unthinking application of any software is very dangerous and care must be taken to assess sampling convergence and stability of results in particular.

All the above methodology is illustrated in a carefully worked example by Hellmich *et al.* (1999), and further applications may be found in Smith *et al.* (1995) and Normand (1999).

7.4 Uncertain or unknown group labels

7.4.1 Introduction

An essential requirement for conducting any of the analyses described in earlier chapters of this book is that the classification score S has been obtained for samples of individuals, each of which has been labeled, and labeled *correctly*, as either N or P. Clearly, the presence of incorrect labeling, or doubt about the accuracy of the labeling, will detract to a greater or lesser extent from the reliability of the analysis, while absence of labeling will rule it out altogether. Very often, however, to achieve correct labeling the sample members may have to be either tested to destruction (if they are inanimate), subjected to some invasive procedure (if they are animate), or only examined in impossible situations such as post mortem. Thus, for example, in an investigation of strengths of materials the sample members need to be subjected to pressure until they break; in study of the detection of breast tumors by mammogram the sample members must provide surgically removed specimens for examination by a pathologist; while the true disease status of individuals presenting global symptoms such as jaundice may only be determinable at post-mortem examination. In some situations (e.g., the breaking-strength example) this is of little consequence (apart perhaps of expense) and one can readily proceed with the labeling and analysis; in others (e.g., the mammogram example), the need to minimize discomfort and danger to the patient may result in use of a less accurate method of labeling; while in extreme cases (e.g., the jaundice example), no labeling is available. In some other applications (e.g., financial ones) there is a delay before the true class is known, so one has at best only partial information.

Since many (but clearly not all) such examples of uncertain or missing labels occur in the fields of medical and epidemiological research, terminology has been borrowed from these fields and a method of labeling that always delivers the correct result is known as a *gold standard*

7.4. UNCERTAIN OR UNKNOWN GROUP LABELS

test or diagnosis. Unfortunately, it is well known that gold standards do not exist for many diseases, and moreover Pfeiffer and Castle (2005) point out that they are in fact rarer than might be thought. As an example they quote research by Guido *et al.* (2003), which estimated only about 60% sensitivity in the detection of cervical cancer of a test that had long been considered to be a gold standard for the disease. So it is clear that methods of ROC analysis should cater for the possibility of sample mislabeling as well as being able to cope with missing labels.

Although many researchers have tackled the problem of uncertain group labeling and absence of a gold standard when the classifier of interest is binary, little progress was made until recently with ROC analysis of continuous classifiers. Henkelman *et al.* (1990) proposed a maximum likelihood method but only for ordinal-scale classifiers; this method was further developed by Beiden *et al.* (2000) using the EM algorithm and including Monte Carlo analysis of uncertainty. Nielsen *et al.* (2002) suggest conducting separate estimation of fn and fp for every classifier threshold and then linking them together to form the ROC curve, but unfortunately the resulting curve is not necessarily monotone and this is a major drawback. So the most relevant developments have been the recent Bayesian ones. Erkanli *et al.* (2006) include an extension to the model described in Section 7.2 above that allows imperfectly labeled groups to be analyzed, while alternative approaches involving parametric estimation and nonparametric estimation have been proposed by Choi *et al.* (2006) and Wang *et al.* (2007) respectively.

Before describing these Bayesian approaches, we note a general framework. As before, the classification score for which the ROC curve is desired is denoted by S. Let L now denote the "group label" variable. In the presence of a gold standard, L is simply deterministic so that the "values" N and P are assigned to sample members definitively and without error. In the presence of a fallible dichotomous test or instrument, L is a binary random variable with some specified (but unknown) probabilities of assigning values P, N respectively to each sample member. However, it may happen that instead of any form of dichotomous test there is only a second continuous classifier U available, so that L has to be derived by dichotomizing U using an appropriate threshold (a situation discussed by Hand *et al.*, 1998). Although we implicitly view the problem in all these cases as that of deriving the ROC for S using L as the reference, in the last case there is a certain ambiguity as

L might equally be derived by thresholding S and the ROC curve then would be for U. To emphasize the symmetry, it is therefore preferable to denote the two continuous classifiers as S_1 and S_2.

The extension to the methods of Section 7.2 provided by Erkanli et al. (2006) is fairly straightforward. L is a binary variable which assigns values P or N to each sample member imperfectly; so assume that the "true" group labels are given by the *latent* (i.e., unobservable) binary variable Z. L and Z can be related to each other in two ways. Either

$$L|Z \sim \text{Bernoulli}(\pi)$$

where $\log\left(\frac{\pi}{1-\pi}\right) = \beta_0$ when $Z =$N and $\beta_0 + \beta_1$ when $Z =$P (the "classical" model); or

$$Z|L \sim \text{Bernoulli}(\zeta)$$

where $\log\left(\frac{\zeta}{1-\zeta}\right) = \beta_0$ when $L =$N and $\beta_0 + \beta_1$ when $L =$P (the "Berkson" model). For full specification the former model also needs $Z \sim$ Bernoulli(ζ) where $\zeta \sim$ Beta(a, b). The parameters β_0, β_1 can be either fixed by the analyst or assigned prior distributions to reflect uncertainty about the relationship between L and Z. The unknown latent variable Z can be simulated from its conditional posterior distribution as an additional step to the Gibbs sampler, but otherwise the process is much as before. Erkanli et al. (2006) give full details of all the steps, and demonstrate the robustness of the method by applying it to gold standard data and comparing the results with those from the analysis of Section 7.2.

The methods of Choi et al. (2006) and Wang et al. (2007) have their own individual features, so are now described in separate sections.

7.4.2 Bayesian methods: parametric estimation

Choi et al. (2006) consider the case of two continuous classifiers in a parametric context. They suppose first that a gold standard exists, so that individuals can be correctly identified as having come from one of the populations P or N. Let S_{1iP}, S_{2iP} be the scores observed on the two classifiers for the ith individual in a random sample of size m from population P, and S_{1jN}, S_{2jN} the scores observed on the same two classifiers for the jth individual in a random sample of size n from population N. The two scores in each population can be put together into a bivariate (column) vector, $\boldsymbol{S}_{iP} = (S_{1iP}, S_{2iP})^T$ (where

7.4. UNCERTAIN OR UNKNOWN GROUP LABELS

the superscript T denotes transpose) and $\boldsymbol{S}_{jN} = (S_{1jN}, S_{2jN})^T$, and the subsequent analysis can be based on these vectors. The starting point is the binormal model so that $\boldsymbol{S}_{iP} \sim N_2(\boldsymbol{\mu}_P, \boldsymbol{\Sigma}_P)$ and $\boldsymbol{S}_{jN} \sim N_2(\boldsymbol{\mu}_N, \boldsymbol{\Sigma}_N)$ where $\boldsymbol{\mu}_P = (\mu_{1P}, \mu_{2P})^T$, $\boldsymbol{\mu}_N = (\mu_{1N}, \mu_{2N})^T$,

$$\boldsymbol{\Sigma}_P = \begin{pmatrix} \sigma^2_{11P} & \sigma_{12P} \\ \sigma_{12P} & \sigma^2_{22P} \end{pmatrix}, \quad \boldsymbol{\Sigma}_N = \begin{pmatrix} \sigma^2_{11N} & \sigma_{12N} \\ \sigma_{12N} & \sigma^2_{22N} \end{pmatrix}.$$

Given the existence of a gold standard, all the usual features of ROC analysis can again be obtained. Since there are two classifiers there will be two possible ROC curves plus allied summary measures. For threshold values $t \in (-\infty, \infty)$ the ROC curves can be constructed by plotting the pairs $\left[1 - \Phi\left(\frac{t - \mu_{kN}}{\sigma_{kkN}}\right), 1 - \Phi\left(\frac{t - \mu_{kD}}{\sigma_{kkD}}\right)\right]$, and the areas under the curves can be calculated as $\mathrm{AUC}_k = \Phi\left(-\frac{\mu_{kN} - \mu_{kP}}{\sqrt{\sigma^2_{kkN} + \sigma^2_{kkP}}}\right)$, for classifier $k = 1, 2$. The difference $\mathrm{AUC}_1 - \mathrm{AUC}_2$ compares the overall accuracy of classifier 1 versus classifier 2, and the closeness of distributions S_N and S_P for classifier k can be given by $\Delta_k = \Phi\left(\frac{\delta_k - \mu_{kP}}{\sigma_{kkP}}\right)$ where δ_k is the 95th percentile of S_{kP} values.

For Bayesian analysis it is more convenient to formulate the bivariate binormal model in terms of the normal marginals for classifier 1 together with the normal conditionals for classifier 2 given classifier 1 values. The necessary conditional means and conditional variances are given from standard theory as

$$E(S_{2iP}|S_{1iP}) = \mu_{2P} + \rho_P(S_{1iP} - \mu_{1P})\frac{\sigma_{22P}}{\sigma_{11P}},$$

$$E(S_{2iN}|S_{1iN}) = \mu_{2N} + \rho_N(S_{1iN} - \mu_{1N})\frac{\sigma_{22N}}{\sigma_{11N}},$$

$$\mathrm{var}(S_{2iP}|S_{1iP}) = \sigma^2_{22P}(1 - \rho_P^2),$$

$$\mathrm{var}(S_{2iN}|S_{1iN}) = \sigma^2_{22N}(1 - \rho_N^2),$$

where

$$\rho_P = \frac{\sigma_{12P}}{\sigma_{11P}\sigma_{22P}} \quad \text{and} \quad \rho_N = \frac{\sigma_{12N}}{\sigma_{11N}\sigma_{22N}}.$$

When prior knowledge is available then informative (e.g., conjugate) priors are appropriate for the various parameters, but in the absence of prior information Choi et al. (2006) suggest independent $N(0, 1000)$ priors for μ_{kP}, μ_{kN}, independently of gamma(0.001, 0.001) priors for $1/\sigma^2_{kkP}, 1/\sigma^2_{kkN}$, and uniform $U(-1, 1)$ priors for ρ_P, ρ_N. They also discuss other possible choices, describe the necessary Gibbs sampling for analysis, and provide Win-BUGS code for implementing the methods.

The move from presence to absence of gold standard requires just a small modification to the above model, similar to that used by Erkanli et al. (2006). Since there is now no information detailing to which population any given individual belongs, let Z again be a latent variable that has this information (so that if Z were observable then Z_j would equal 1 if the jth individual came from P and 0 if it came from N) and let m be the number of individuals observed. We now assume that $Z_j \sim$ Bernoulli (π), i.e., that $\Pr(Z_j = 1) = 1 - \Pr(Z_j = 0) = \pi$, so that the probability density for the bivariate classification score of the jth individual is $p(\cdot)^{Z_j} g(\cdot)^{1-Z_j}$ where $p(\cdot), g(\cdot)$ are $N_2(\boldsymbol{\mu}_P, \boldsymbol{\Sigma}_P)$ and $N_2(\boldsymbol{\mu}_N, \boldsymbol{\Sigma}_N)$ densities respectively. Inferences are the same as when a gold standard is available except that the prevalence of each population P and N must now be estimated. The additional parameter π also needs a prior distribution, and Choi et al. (2006) suggest either a Beta (1, 1) or a Beta (0.5, 0.5) distribution in the absence of prior information. Once again they provide the necessary Win-BUGS code for implementing the analysis.

To illustrate the methodology, Choi et al. (2006) analyzed test scores for Johne's disease in dairy cattle. Two enzyme-linked immunosorbent assay (ELISA) tests are commercially available in the United States for detecting serum antibodies to the cause of this disease; so these tests were used to serologically test cows in dairy herds of known infection status. The data set comprised 481 cows; in fact, a gold standard test was also available and applied, yielding 88 infected (population P) and 393 noninfected (population N) cows, and so the two ELISA tests could be compared either using or ignoring the gold standard information.

Initial data analysis showed that log-transformed scores of both tests had only minor departures from normality; so a binormal model was deemed to be perfectly acceptable. Parameter estimates when ignoring the gold standard information (NGS) were in fact quite similar to their counterparts when the gold standard information was used (GS), in all cases falling well within the 95% probability intervals of the latter. The authors surmised that the agreement was probably caused by the reasonable separation between the classification scores in the two populations for one of the tests. The estimated prevalence $\hat{\pi}$ of the disease in the NGS case was 0.18 with a probability interval $(0.14, 0.23)$, while the sample proportion infected using the GS information was $88 \div 481 = 0.18$. The GS and NGS ROC curves were very close for each test, and their AUCs were likewise very similar: 0.93 vs 0.95 for one test, 0.96 vs 0.97 for the other. Using the GS analysis, the overlaps Δ_k between populations P and N were estimated to be just 8% for one test and 24% for the other. This clear separation was probably caused by only including the most severely infected cows according to the GS test; in typical practice when more mildly infected cows are included in P, the overlap is expected to be in excess of 70%.

7.4.3 Bayesian methods: nonparametric estimation

Wang *et al.* (2007) provided a nonparametric approach to the analysis of a continuous classifier score S in the case where the group label variable L is fallible. A full description of the methodology involves extensive and fairly intricate notation, so we restrict ourselves here to a somewhat simplified summary.

We assume as usual that higher values of S indicate population P, so that for a given threshold t the classifier allocates an individual to P if its S value exceeds t. The ROC curve is formed by plotting the (fp, tp) values for a range of thresholds t, so let us assume that K thresholds $t_1 < t_2 < \ldots < t_K$ have been selected and that the classifier has been applied to a sample of data at each threshold. Let $\alpha^{(i)}, \beta^{(i)}$ be the true (unknown) $fp, 1 - tp$ values at threshold t_i for $i = 1, \ldots, K$. Since the thresholds are ordered it follows that $\alpha^{(1)} \geq \alpha^{(2)} \geq \ldots \geq \alpha^{(K)}$ and $\beta^{(1)} \leq \beta^{(2)} \leq \ldots \leq \beta^{(K)}$. Define boundary values $\alpha^{(K+1)} = \beta^{(0)} = 0$ and $\alpha^{(0)} = \beta^{(K+1)} = 0$. Moreover, since the group label variable L is fallible, let its true (unknown) false positive and false negative rates be $\alpha = \Pr(P|N), \beta = \Pr(N|P)$ respectively. Finally, let θ be the true (unknown) prevalence of population P so that θ is the probability that any individual in the group being sampled actually belongs to P (and hence $1 - \theta$ is the probability that it belongs to N). This completes the set of unknown parameters in the problem.

Now turn to the data. Suppose that n individuals are in the sample, and that n_P, n_N of them are labeled by L as belonging to populations P and N respectively. Considering the classifier scores S at each of the K thresholds, let x_{iP}, x_{iN} denote the number of individuals labeled P, N respectively by L that are classified P at threshold t_{i-1} but N at threshold t_i for $i = 2, \ldots, K - 1$. At the boundaries we have special cases: $x_{1P}, x_{1,N}$ denote the number of individuals labeled P, N respectively that are classified N by S, and x_{KP}, x_{KN} denote the number of individuals labeled P, N respectively that are classified P by S. By deriving the probability of each occurrence in terms of the model parameters, the likelihood of the sample can be shown to have the *ordered multinomial* form

$$\mathcal{L} \propto \prod_{i=1}^{K+1} [\theta(\beta^{(i)} - \beta^{(i-1)})(1-\beta) + (1-\theta)(\alpha^{(i-1)} - \alpha^{(i)})\alpha]^{x_{iP}} \times \\ [\theta(\beta^{(i)} - \beta^{(i-1)})\beta + (1-\theta)(\alpha^{(i-1)} - \alpha^{(i)})(1-\alpha)]^{x_{iN}}.$$

However, a basic problem with this model is that there are $2K+3$ parameters (the $\alpha^{(\cdot)}, \beta^{(\cdot)}, \alpha, \beta$, and θ), but only $2K+1$ degrees of freedom for estimation. Hence the model is unidentifiable and not all parameters can be estimated. This problem can be overcome if there are different groups of individuals that have different prevalences $\theta_1, \theta_2, \ldots, \theta_G$ of population P, but such that the behavior of S (with respect to true/false positives) is the same in each group. The latter condition is usually deemed to be reasonable, so if there are G such groups and independent samples are obtained in each of them then the likelihood of the full set of data is just the product of all the above expressions evaluated for each separate group. Now there are $2K+G+2$ unknown parameters and $G(2K+1)$ degrees of freedom, so the model is identifiable whenever $G \geq 2$.

The one extra consideration is to guarantee monotonicity of the ROC curve, and for this we need to constrain all the $\alpha^{(\cdot)}, \beta^{(\cdot)}$ to be positive. Wang et al. (2007) ensure this is the case by taking a Bayesian approach with Dirichlet priors for these parameters, and they additionally suggest various Beta prior distributions for the remaining parameters α, β, θ_g. Multiplying these priors by the data likelihood produces the posterior distribution, and Markov-chain Monte Carlo is then employed to obtain posterior estimates of the parameters and hence the ROC curve and its summaries. However, it is not straightforward to construct the necessary conditional distributions for the Gibbs sampler directly, so Wang et al. (2007) tackle the problem via data augmentation as described by Tanner and Wong (1987). They derive expressions for the full conditional probabilities used in the Gibbs sampler, and offer their MATLAB® software to interested parties upon request.

To illustrate their methodology, Wang et al. (2007) also use data from tests for Johne's disease in cows, but not from the same study as that investigated by Choi et al. (2006). In their case there was one

ELISA (continuous) test and a fecal culture (FC) binary test. The data comprised observations on 2,662 cows from 55 herds, being taken from a multipurpose study on infectious disease that had previously been conducted in Denmark. The same study formed the basis of analysis conducted by Nielsen *et al.* (2002); so the datasets in the two papers were similar. The division into two groups, the minimum required by the Bayesian methodology, was done by a veterinarian using the same criterion as that adopted by Nielsen *et al.* (2002), whose frequentist method of separate ML estimates also requires division of the data into several groups exhibiting different population prevalences. The accumulated evidence from these studies suggests that the different prevalence assumption was reasonable.

Noninformative priors were used in the Bayesian analysis, and four different models were implemented according to the number K of thresholds used to form the classification data: $K = 10, 20, 50$, and 100. The four resulting Bayesian ROC curves were all very similar to each other, and close in position to the frequentist one produced by the method of separate ML estimates. The latter, however, was not monotone, a failing already commented on earlier in the chapter, while the Bayesian methodology ensured monotone curves in all cases. Increasing K captures more detail on the ROC but introduces extra parameters, makes the model more sensitive to the prior information, and increases the running time of the program. The AUC values were $0.784, 0.785, 0.758, 0.748$ for $K = 10, 20, 50, 100$ respectively; plots of other parameter estimates and their credible intervals are also provided by Wang *et al.* (2007).

7.5 Further reading

For readers unfamiliar with the Bayesian approach, a good general text is the one by Lee (2004); modern Bayesian methods for classification and regression are surveyed by Denison *et al.* (2002), and Markov-chain Monte Carlo methods are discussed by Brooks (1998).

Related work to that above on absence of a gold standard, but in the context of binary or nominal variables, may be found in the papers by Enøe *et al.* (2000), Garrett *et al.* (2002), and Branscum *et al.* (2005), and further references from these papers may also be of interest.

Chapter 8

Beyond the basics

8.1 Introduction

Basic ROC curves summarize the score distributions of objects from two classes. Situations requiring such summaries are very common—and hence the widespread use of ROC curves and the statistics derived from them. However, in many situations things are more complicated than this basic notion. Clearly such further complications can arise in an unlimited number of ways, so we cannot examine all of them. Instead, in this chapter, we describe some of the work on just a few of the most important such generalizations.

8.2 Alternatives to ROC curves

ROC curves portray the performance of classification rules by plotting the true positive rate against the false positive rate as the classification threshold is varied, but there are other ways of displaying the performance of a classifier. Under appropriate circumstances, or when one is seeking to understand or communicate particular aspects of classifier behavior, these other approaches might be more suitable than ROC curves. Typically, however, they are simply alternative ways of representing the same information. This section describes some of these alternative approaches.

The most trivial alternative is to reverse an axis. The standard form of the ROC curve plots the true positive rate, $p(s>t|P)$, as the ordinate, and the false positive rate, $p(s>t|N)$, as the abscissa. As noted above, and as is apparent when one comes to extend the

ROC curve to more than two classes, this represents an asymmetric treatment of the two classes. One alternative would be to plot true positive against true negative by reversing the horizontal axis. Another would be to plot false negative against false positive by reversing the vertical axis. But other, less trivial, alternative representations are also possible, and some of these are now considered.

A common variant is the *Lorenz curve*, described in Chapter 2. The traditional form of this curve, as widely used in economics, plots $Y(t) = p(s \leq t|P)$ on the vertical axis against $X(t) = p(s \leq t)$ on the horizontal axis. If we denote the traditional ROC curve axes by $y(t) = p(s > t|P)$ and $x(t) = p(s > t|N)$, then the Lorenz curve axes are obtained by the simple linear transformations $Y = 1 - y$ and $X = (1-y) p(P) + (1-x) p(N)$. If the distributions of the scores in classes P and N are identical, then so also will be the distributions of the scores in class P and the overall mixture distribution of scores, so that the Lorenz curve will lie along the diagonal.

A very closely related version, widely used in business and commercial applications, is the *lift curve*. In a lift curve the cumulative proportion classified as class P is plotted against the cumulative proportion in the population: $p(P|s \leq t)$ vs $p(s \leq t)$. Once again, if the classifier is poor at separating the classes, the two sets of values will be very similar, so that a straight diagonal line will be plotted. If the classifier is very effective, then the curve will tend to be low for most values of $p(s \leq t)$, before swinging sharply up towards the end.

In Chapter 1, we defined the *precision* and *recall* of a classifier, these being terms widely used in the information retrieval and webmining literature. In our terms, precision is the positive predictive value, $p(P|s > t)$, and recall is the true positive proportion, $p(s > t|P)$. *Precision-recall* curves plot precision on the vertical axis against recall on the horizontal axis.

Clearly, precision-recall curves can be obtained from ROC curves by first transforming the horizontal axis:

$$(p(s > t|N), p(s > t|P)) \Rightarrow \left(1 - \frac{p(s > t|N) p(N)}{p(s > t)}, p(s > t|P)\right)$$
$$= (p(P|s > t), p(s > t|P));$$

and then interchanging the axes

$$(p(P|s > t), p(s > t|P)) \Rightarrow (p(s > t|P), p(P|s > t)),$$

8.2. ALTERNATIVES TO ROC CURVES

so in some sense they add nothing. However, they come into their own when the classes are severely unbalanced, with one class being very small compared to the other.

To see the difficulties which can arise in such unbalanced situations, recall the situation described in Section 1.3, in which we supposed we had a classifier that correctly classified 99% of the class N cases and 99% of the class P cases. An ROC plot of true positives, $p(s > t|P)$, against false positives, $p(s > t|N)$, would consist of a line connecting $(0,0)$ to $(0.01, 0.99)$ and a line connecting $(0.01, 0.99)$ to $(1, 1)$, suggesting the classifier was superb. But, as in Section 1.3, now suppose that in fact class P is very rare, with only 1 in a thousand of the population being in class P. We saw there that only 9% of the cases the classifier assigned to class P would in fact belong to class P. The classifier, which appears to perform so well when looked at from the true positive and false positive perspective of the ROC curve, no longer seems so attractive. The precision-recall plot draws attention to this anomaly, by explicitly displaying, on the vertical axis, the fact that, in this example, only a small proportion of those assigned to the P class really are from that class.

Such situations with severely unbalanced class sizes are quite common. For example, they arise in screening for rare diseases, in fraud detection, and when seeking rare events or particles in high energy physics experiments.

In Chapter 2 we defined $c(P|N)$ to be the cost incurred by misclassifying a class N point as class P, and $c(N|P)$ to be the cost of misclassifying a class P point as class N. Drummond and Holte (2000) included costs explicitly when developing an alternative representation which they call *cost curves*. The horizontal axis is

$$H = \frac{p(P) c(s \leq t|P)}{p(P) c(s \leq t|P) + p(N) c(s > t|N)} \quad (1)$$

which they call the *probability cost function*, and the vertical axis is

$$V = \frac{p(s \leq t|P) p(P) c(s \leq t|P) + p(s > t|N) p(N) c(s > t|N)}{p(P) c(s \leq t|P) + p(N) c(s > t|N)}. \quad (2)$$

For a given problem $p(P)$ and $p(N) = 1 - p(P)$ are fixed, so that (1) is a simple monotonic transformation of the ratio of misclassification costs $c(s \leq t|P)/c(s > t|N)$. The numerator in (2) is the expected cost, arising from the proportion of class P misclassified as class N and the proportion of class N misclassified as class P. The denominator in (2) is just a normalizing factor, being the maximum value that the numerator can take (when $p(s \leq t|\text{P}) = 1$ and $p(s > t|\text{N}) = 1$).

Substituting (1) into (2) gives

$$V = Hp(s \leq t|\text{P}) + (1 - H)p(s > t|\text{N}),$$

so that, if we fix the classification threshold t (corresponding to a single point on a standard ROC curve), a plot of V against H is a straight line.

The particular merit of this representation of a classifier's performance is that, at a given value of H, the vertical distance between the straight lines corresponding to two classifiers gives the difference in expected cost of the classifiers for that H value.

In the case in which the two misclassification costs are equal, (1) reduces to $H = p(\text{P})$, so that (2) becomes simply the error rate of the classifier. The vertical distance between two classifiers at a given value of H is then just the error rate difference at that H.

Two misclassification costs appear in (1) but in fact they involve only a single degree of freedom. We have

$$\begin{aligned} H &= \frac{p(\text{P}) c(s \leq t|\text{P})}{p(\text{P}) c(s \leq t|\text{P}) + p(\text{N}) c(s > t|\text{N})} \\ &= \frac{p(\text{P})}{p(\text{P}) + p(\text{N}) \{c(s > t|\text{N})/c(s \leq t|\text{P})\}} \\ &= \frac{p(\text{P})}{p(\text{P}) + p(\text{N}) k}, \end{aligned}$$

where $k = c(s > t|\text{N})/c(s \leq t|\text{P})$.

Adams and Hand (1999) built on this feature, and considered plots of the misclassification cost of a classifier against $(1+k)^{-1}$ (the transformation meaning that the horizontal axis ranges from 0 to 1, instead of 0 to ∞, as would be the case if we used the raw ratio k). Their main concern was to compare classification rules, and, noting the intrinsic noncommensurability of different cost pairs $(c(s \leq t|\text{P}), c(s > t|\text{N}))$ (since only the ratio counts), reduced things to simply summarizing the sign of the superiority of different classifiers at different cost ratios, rather than the (noncommensurate) magnitude. This led to the LC index for comparing classifiers that was described in Section 3.5, based on aggregating over a distribution of likely k values.

Yet another variant is the *skill plot* (see Briggs and Zaretzki, 2008). The expected misclassification cost is

$$p(s \leq t|\text{P})\, p(\text{P})\, c(s \leq t|\text{P}) + p(s > t|\text{N})\, p(\text{N})\, c(s > t|\text{N}),$$

the numerator in (2). Whereas Drummond and Holte (2000) standardized the expected cost by dividing by the maximum value that the expected cost could take, $p(\text{P})\, c(s \leq t|\text{P}) + p(\text{N})\, c(s > t|\text{N})$, Briggs and Zaretzki's method is equivalent to standardizing by $p(\text{P})\, c(s \leq t|\text{P})$ or by $p(\text{N})\, c(s > t|\text{N})$ according as $p(\text{N})$ is respectively less than or greater than $c(s > t|\text{N})$, and then inverting the interpretation by subtracting the result from 1. The standardization here is based on the observation that if $p(\text{N}) < c(s > t|\text{N})$ then the optimal naive approach is to classify everything into class N, leading to an expected cost of $p(\text{P})\, c(s \leq t|\text{P})$, while if $p(\text{N}) > c(s > t|\text{N})$ then the optimal naive approach is to classify everything into class P, leading to an expected cost of $p(\text{N})\, c(s > t|\text{N})$.

Yet other variants are possible, of course, but those above are the ones that have received most attention in the literature.

8.3 Convex hull ROC curves

The ROC curve shows the true positive rate plotted against the false positive rate. When the ROC curves for two classifiers cross, it means that one of them has a superior true positive rate for some values of the false positive rate, while the other has a superior true positive rate for other values of the false positive rate. That is, that neither classifier dominates the other, so that neither is uniformly superior: which classifier is best depends on the circumstances of its use.

However, Provost and Fawcett (1997, 1998) and Scott *et al.* (1998) have presented a very elegant way of combining classifiers based on combining ROC curves. This is illustrated in Figure 8.1, which shows two ROC curves that cross. For false positive rate $fp3$, the true positive rate for classifier 1 is labeled as $tp1$, and that for classifier 2 as $tp2$. That is, it would seem that if we wanted to achieve a false positive rate of value $fp3$, we would choose classifier 2, since it has the larger true positive rate. However, the same false positive rate could be achieved by randomly choosing between classifier 1 with its classification threshold chosen so that its false positive rate is $fp1$, and classifier 2 with its classification threshold chosen so that its false positive rate is $fp2$, in

Figure 8.1: Convex hull of two ROC curves

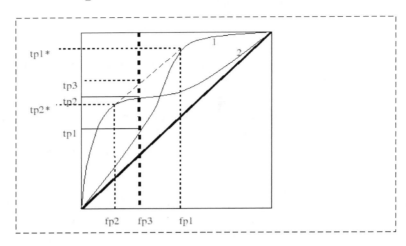

such proportion that the mean false positive rate was $fp3$. That would then yield a true positive rate which was a similarly weighted mean of $tp1*$ and $tp2*$, namely $tp3$. By randomly selecting the classifiers in this way we have achieved a combined classifier which has superior performance to either one separately.

The ROC curve for the combined classifier, namely the path traced out by $tp3$, is the *convex hull* of the separate ROC curves. As an aside, a few words of explanation about terminology might be appropriate here. A function f is said to be *concave* on an interval $[a, b]$ if for every pair of points x_i and x_j in $[a, b]$, and for any $0 < \lambda < 1$

$$f\left[\lambda x_i + (1-\lambda) x_j\right] \geq \lambda f(x_i) + (1-\lambda) f(x_j).$$

Thus one might have expected the curves above to be referred to as *concave* rather than convex. The term convex has arisen from the definition of a convex hull as the boundary of a set of points $x_1, ..., x_n$, defined as

$$\left\{\sum_{i=1}^n \lambda_i x_i : \lambda_i \geq 0 \ \forall i \ \text{and} \ \sum_{i=1}^n \lambda_i = 1\right\}.$$

The use of the word "convex" to describe the upper bounding concave function of a family of ROC curves is thus not really appropriate—however, it has become standard usage.

Figure 8.2: Improving a single ROC curve by the convex hull method

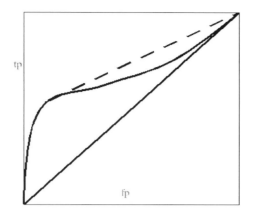

The principle behind this method can also be used to improve classification performance for individual classifiers, as shown in Figure 8.2.

8.4 ROC curves for more than two classes

Standard ROC curves represent the trade-off, in the two-class case, between the proportion of class P points correctly classified and the proportion of class N points misclassified, as the classification threshold varies. They do this by plotting the proportion of class P points correctly classified, on the vertical axis, against the proportion of class N points incorrectly classified, on the horizontal axis. This means that ROC curves lie in a two-dimensional space. When more than two classes are involved, we enter the realm of needing to display the relationships between more than two variables simultaneously, so that things naturally become more complicated. Moreover, when there are more than two classes, there is some flexibility about exactly what aspects of the classification problem are of interest. Different researchers may be concerned with different aspects. This is a deeper matter than simply deciding which of various equivalent representations of the ROC curve to choose, of the kind discussed in Section 8.2. It has the consequence that different researchers have proposed different ways of generalizing the ROC curve. In what follows, for convenience, we will denote the number of classes by c and index these classes by the integers $1, ..., c$.

As we have already seen, the two-class classification situation is the most common, since two classes fit many naturally occurring problems: yes/no, right/wrong, sick/healthy, etc. In general, moreover, one can specify a two-class problem by defining one class, and letting the other class simply be its complement. But the special case of three classes is also often of interest, and frequently the three-class case arises as a derivative of the two-class case. For example, it is not uncommon to have two well-defined classes, along with a third class of objects about whose membership one is not certain. Thus, in credit scoring in the retail financial services sector, one might define bad customers as those who have fallen three months behind with repayments, and good customers as those who have never fallen behind. This leaves a so-called indeterminate third class between good and bad. In medicine, a radiologist might have to distinguish between a benign tumor, a malignant tumor, and some other cause. For reasons such as this, although the two-class case (and the simple two-dimensional associated ROC curves) are by far the most widely used and intensively investigated cases, the three-class case also has a literature of its own.

The conventional ROC display, as followed in this book, is to plot the true positive rate on the vertical axis and the false positive rate on the horizontal axis. One alternative, however, is to plot the true positive rate on the vertical axis and the true negative rate (that is, $1 -$ (false positive rate)) on the horizontal axis. This results in a reflection of the traditional ROC curve about a vertical axis. All discussion and interpretation follows through, though now adjusted for the reflection.

The advantage of this simple alternative representation is that it permits immediate generalization to multiple classes. We can define a ROC hypersurface for c classes in terms of c axes, with the ith axis giving the proportion of the ith class correctly classified into the ith class. (In the two-class case, the true negative rate is the proportion of class N correctly classified.) A point on the surface simply tells us that there is a configuration of the multiclass classifier which will yield the indicated correct classification rates for each of the classes.

When $c = 3$, this approach means we have to display just three axes, and there are various well known methods for such situations, including contour plots and 3-dimensional projections into 2-dimensional space. However, when $c > 3$ the situation defies convenient graphical representation.

8.4. ROC CURVES FOR MORE THAN TWO CLASSES

The extension to more than two classes just described has at its heart a simplification. With c classes, there are in fact $c(c-1)$ different possible types of misclassification. The multidimensional ROC curve extension described above does not reflect this variety, but merely categorizes classifications into correct and incorrect for each class, not distinguishing between the different kinds of incorrect assignment. That is, it condenses the conventional confusion matrix (a cross-classification of true class by predicted class for a sample of objects) from c^2 cells (that is, the $c(c-1)$ cells of misclassifications plus the c cells of correct classifications) into just $2c$ values.

Clearly the results of any multi-class classifier can be condensed in this way (at the cost of loss of information about the classifier performance, of course), but such a situation also arises from an important special case. This is when all objects are scored on a single scale and the values of this scale are divided into c intervals, each interval corresponding to prediction into one of the classes. The proportion in each interval which actually comes from the class into which the interval predicts serves as one of the axes, one for each class. Since, to define such intervals requires just $(c-1)$ thresholds, varying the thresholds traces out a $(c-1)$-dimensional hypersurface—the multidimensional ROC hypersurface in the c dimensional situation described above.

In some situations one may be uneasy about analyzing the multiclass case in terms of simple correct/incorrect classifications, but may also want to consider the nature of the incorrect classifications. For example, a mistaken diagnosis can be of greater or lesser severity, according to the disease to which the case is incorrectly assigned, and the treatment, or lack of treatment, which follows. In this case, one can define a generalization of the ROC curve as being based on $c(c-1)$ axes, each corresponding to one of the possible kinds of misclassification. Of course, this approach defies simple graphical representation of the ROC curve even for c as small as 3, since then the ROC curve is a 5-dimensional surface lying in a 6-dimensional space. Note also that, with axes plotted in terms of misclassification rates rather than correct classification rates, this form of curve is inverted.

These two generalizations of ROC curves can also be described in terms of the costs associated with the misclassifications. Varying these costs is equivalent to varying the classification thresholds, and so tracing out the ROC hypersurface (or curve when $c = 2$). Taking the simpler ROC curve generalization first (where we consider merely correct and incorrect classifications), in these terms an object from class i has an associated cost, γ_i, if it is misclassified, and each vector of costs $(\gamma_1, \gamma_2, ..., \gamma_c)$ corresponds to an operating point on the ROC hypersurface (in fact we can normalize by setting one of the costs to 1). The more elaborate generalization which allows for the fact that there are different kinds of misclassification requires an entire $c \times c$ matrix of costs. If correct classifications are taken to incur zero cost, then there are $c(c-1)$ costs to be chosen, and varying this operating point traces out the ROC hypersurface (and, as before, one of the costs can be set to 1).

We have thus described two extensions of the simple ROC curve to more than two classes. The first approach is equivalent to regarding each misclassification from a given class as equally costly, regardless of the class into which the object has been misclassified. That is, each of the true classes has a unique common cost associated with any misclassifications of its members. The ROC surface is traced out by letting these common cost values vary. In contrast, the more general approach regards each misclassification as having its own cost. There are $c(c-1)$ of these costs, since members of each of the c classes can be misclassified into $(c-1)$ other classes. The ROC surface in this case is obtained by letting each of the $c(c-1)$ cost values vary. Landgrebe and Duin (2007, 2008) describe an interesting intermediate generalization between these two. They suppose that the off-diagonal elements of the cost matrix are proportional to a given vector, with the constant of proportionality differing between rows and being allowed to vary. Thus, like the simple generalization, it yields a ROC curve in a c-dimensional space, but the initial choice of misclassification cost matrix allows the costs to depend on the class to which the misclassification is made.

The two types of generalization of the two-class ROC curve to more than two classes lead to two kinds of generalization of the AUC.

The first extension of the ROC curve described above used, as the ith axis, the proportion of class i correctly classified into class i. When there are just two classes, the area under the ROC curve (that is, the AUC) is the area of the unit square lying between the origin and the ROC curve. For c classes, this immediately generalizes to the volume between the origin and the ROC hypersurface. For the two-class case, one intuitive interpretation of the AUC is that, if one randomly (uniformly) chooses what proportion of the N class to classify correctly, then the AUC is the mean proportion of class P individuals which

8.4. ROC CURVES FOR MORE THAN TWO CLASSES

are correctly classified. This interpretation generalizes to the case of c classes: if one randomly (uniformly) chooses what proportions of classes 1 to $c-1$ one wishes to classify correctly (subject to it being possible to achieve such a set of proportions), then the hypervolume under the ROC hypersurface is the average proportion of class c individuals which are correctly classified.

With just two classes, a classification rule will yield estimated probabilities of belonging to each of the two classes. For an object with descriptive vector x, we will have estimates $p(\text{P}|x)$ and $p(\text{N}|x)$. Since $p(\text{P}|x) + p(\text{N}|x) = 1$, these estimates can clearly be graphically represented as points on the unit interval. For three classes, labeled 1, 2, and 3, each object will give a triple of estimated class memberships $(p(1|x), p(2|x), p(3|x))$. Now, since $p(3|x) = 1 - p(1|x) - p(2|x)$, we can represent these triples as points in two dimensions. However, rather than arbitrarily choosing two of the three $p(i|x)$ values to serve as axes, a display developed in the compositional analysis literature can be adopted. An equilateral triangle of unit height is defined, with the vertices representing the three triples $(1,0,0)$, $(0,1,0)$, and $(0,0,1)$. That is, the vertex labeled $(1,0,0)$ corresponds to certainty that the object belongs to class 1, and so on. For any point within the triangle, the sum of the three perpendicular distances from the point to the sides is 1. Since, for any point within the triangle, all three of these distances must be nonnegative, they can represent probabilities. We thus see that, roughly speaking, the closer a point is to a vertex the further is its perpendicular distance to the opposite edge of the triangle, and so the greater is its probability of belonging to the class represented by that vertex. These triangles are called *ternary* diagrams or *barycentric* coordinate spaces. An illustration is given in Figure 8.3.

In Section 2.4, we saw that another interpretation of the AUC was that it was the probability that, given a randomly chosen pair of objects, one from each class, the classifier will give a higher score to the class P object than to the class N object. This idea can be generalized to more than two classes. For example, suppose that we have three objects, with the classifier yielding the triple probability estimate (p_{i1}, p_{i2}, p_{i3}) for the ith of these objects. Assign one of the three objects to each of classes 1, 2, and 3. There are 6 possible ways this assignment could be made. Let the class to which the ith object is assigned be denoted by (i). Now, choose that assignment which maximizes $\sum_{i=1}^{3} p_{i(i)}$. Then the probability that this assignment is the correct one is equal

Figure 8.3: A ternary diagram representing the probabilities that each data point belongs to each of three classes. The distance from the edges of the triangle is shown for one of the points. These distances represent the probabilities that the point belongs to the class of the vertex opposite the edge.

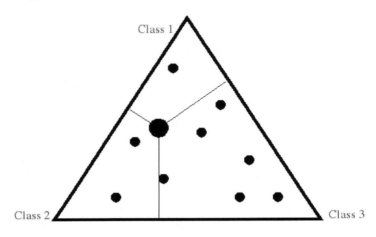

to the volume under the ROC surface (the VUS). Note that, if the assignment is random, then the probability that the assignment of the three objects to the three classes is correct is $1/6$, and this can also be seen to be the VUS for a chance three-class ROC surface (using the approach based on the three "proportion correct" axes), which is a plane connecting the the three points $(1,0,0)$, $(0,1,0)$, and $(0,0,1)$. For the special case in which the objects are scored on a single scale, the VUS gives the probability that the three objects will be placed in the correct order.

Several authors have also examined generalizations of the AUC when the generalized ROC curve is based on the different kinds of misclassification. For example, Everson and Fieldsend (2006) propose using the volume of the $c(c-1)$-dimensional hypercube that is dominated by elements of a classifier's ROC hypersurface. They also derive the expected volume under random classification. In particular, when the ROC hypersurface lies in a hypercube defined by the $c(c-1)$ different types of misclassification as axes, random allocation produces a surface which is the simplex in this $c(c-1)$-dimensional space, with

8.4. ROC CURVES FOR MORE THAN TWO CLASSES

vertices at a distance $(c-1)$ from the origin on each axis. This leads to the volume not dominated by the random classifier to be

$$\frac{(c-1)^{c(c-1)}}{[c(c-1)]!} - \frac{c(c-1)(c-2)^{c(c-1)}}{[c(c-1)]!}.$$

Note that, when $c = 2$, this reduces to $1/2$, as expected. It also decreases rapidly as c increases. Everson and Fieldsend (2006) propose a generalization of the Gini coefficient based on this measure.

A different approach to generalizing the AUC has been developed by Hand and Till (2001), though again focusing on the pairwise comparisons between all the classes. To illustrate, suppose that a classification rule has yielded estimates $\hat{p}(i|x)$ of the probabilities that an object with measurement vector x belongs to class i for $i = 1, ..., c$. That means that there are c alternative scales from which to choose as the basic scale on which to calculate the AUCs, and Hand and Till propose tackling the problem by focusing attention on $\hat{p}(i|x)$ and $\hat{p}(j|x)$ when comparing classes i and j. That is, the AUC between classes i and j is estimated using the scores $\hat{p}(i|x)$ and $\hat{p}(j|x)$. Now, when there are only two classes, $\hat{p}(1|x) = 1 - \hat{p}(2|x)$, so that, provided a consistent order is adopted, the AUCs calculated using the two scales yield identical results. When there are more than two classes, however, this will generally not be the case. Hand and Till thus propose using, as a measure of separability between the two classes i and j, the average of the AUCs for the two classes calculated using the scales $\hat{p}(i|x)$ and $\hat{p}(j|x)$. Define $A(i|j)$ to be the probability that a randomly drawn member of class j will have a lower estimated probability of belonging to class i than a randomly drawn member of class i. Then $A(i|j)$ is the AUC measuring the separability between the scores for classes i and j when these scores are given by $\hat{p}(i|x)$. Similarly, we define $A(j|i)$ to be the AUC between the distributions of $\hat{p}(j|x)$ for classes i and j. The separability between classes i and j is then defined to be

$$\bar{A}(i,j) = (A(i|j) + A(j|i))/2.$$

The overall measure is then the average of these comparisons over all pairs of classes:

$$M = \frac{2}{c(c-1)} \sum_{i<j} \bar{A}(i,j).$$

An alternative (and probably less useful) approach would be to estimate classification rules separately for each pair of classes, so that it became a sequence of two-class comparisons. This would overcome the problem of the multiplicity of potential scales to use for each pairwise comparison, but it is less appealing because the single c-class classification rule (which produced the c estimates $\hat{p}(i|x)$ above) has been replaced by $c(c-1)/2$ separate classification rules.

One of the merits of aggregating the results of pairwise comparisons is that, like the two class AUC, it is independent of class priors. Some other approaches (e.g., aggregating comparions of each class with the mixture distribution of all other classes) will have to contend with issues of priors.

8.5 Other issues

As noted at the beginning of this chapter, ROC curves and their properties have been explored in a large number of special situations, for which variants of the basic form are appropriate, or for which particular aspects of the curve are of principle interest. We have described some of these aspects in the preceding sections. Others include the following.

(i) Estimating classifier parameters by maximizing the AUC.
The area under an ROC curve, the AUC, is a convenient summary of the performance of a classifier. However, when estimating the parameters of classification rules or conducting a model search over a class of classifiers, it is common to use other measures of performance or goodness-of-fit between data and model. For example, in statistics the likelihood is a very popular measure and in machine learning error rate is a popular choice. Now clearly, different measures are likely to lead to different models being chosen. With this in mind, several authors have proposed using the AUC directly to estimate parameter values (e.g., Yan *et al.*, 2003; Toh *et al.*, 2008)

(ii) AUC measures for nonbinary outcomes.
The AUC is the probability that a randomly chosen member of class P will have a score larger than a randomly chosen member of class N. As we saw in Chapter 3, we can estimate this probability by considering all pairs of objects, one object from the class P sample and one from the class N sample, and calculating the proportion of pairs for which the class P objects have a larger score than the class N object. This can easily be extended to

cope with ties. If we give a value 1 to the class P objects and a value 0 to the class N objects, then the AUC is seen simply to be a measure of concordance between the orderings produced by the classification and the score. This perspective permits immediately generalization to nonbinary outcomes. For example, if the outcome is on a continuous scale, we obtain a rank order correlation coefficient.

(iii) Time and ROC curves.
In many situations, the true class is an outcome indicator which is not determined until long after the score has been calculated. For example, this is the case in disease prognosis or in credit default, where one is trying to predict a future outcome. Indeed, in some cases, the true class may not actually exist at the time the score is calculated. In stationary situations this poses no difficulties: the past distributions from which the classifiers are derived, and the score distributions calculated, will also apply in the future. However, many situations are nonstationary, and then difficulties do arise. For example, a classifier used by a bank for determining if a potential customer was a good or bad risk will be based on past data whose distributions may be quite unlike those of the present.

There is no ideal solution to this problem, though various approaches have been tried, including replacing the outcome classification by a proxy which can be observed sooner (so that the past data used to construct the classifier is from the more recent past).

Another issue involving time is that sometimes the predictive power of a score based on data coming from closer to the final outcome is superior to that of a score based on earlier data. In a medical context, this would mean that tests made closer to the time that symptoms became apparent may be more predictive. That is, the performance of the classification rule would depend on when the predictor variables were measured.

8.6 Further reading

Variants of ROC curves have been described by a number of authors: Adams and Hand (1999), Drummond and Holte (2000), Davis and

Goadrich (2006), Briggs and Zaretski (2008), and others. There seems to be something about ROC analysis that leads to the occasional reinvention of the methods: we speculate that it is partly because of the apparent simplicity of the basic problem, with very few degrees of freedom having to be represented, and partly because the need for such representations are so ubiquitous, occurring in widely different intellectual communities which do not (and cannot be expected to) read each other's literature.

The convex hull idea for combining classifiers was described by Provost and Fawcett (1997, 1998), and Scott *et al.* (1998), although there are also traces of it in earlier literature.

The version of the multiple ROC curve based on axes that give the proportion of each class correctly classified has been explored in depth for the particular case of three classes by various authors, including Mossman (1999), Dreiseitl *et al.* (2000), Nakas and Yiannoutsos (2004), and He *et al.* (2006). The more general perspective, in which the proportions from each class misclassified into other classes serve as the axes, has been explored by authors including Hand and Till (2001), Edwards *et al.* (2004, 2005), Ferri *et al.* (2004), and Everson and Fieldsend (2006). The intermediate perspective in which an arbitrary cost matrix serves as the starting point, but where each row can be rescaled was introduced by Landgrebe and Duin (2007, 2008).

Dreiseitl *et al.* (2000) derived a form for the standard deviation of the VUS in the three class case. Everson and Fieldsend (2006) and Hand and Till (2001) are amongst authors who have extended the AUC to more than three classes.

Some multiclass work has also looked at the special case in which the scores for the objects lie on a single continuum—see, for example, Nakas and Yiannoutsos (2004).

Chapter 9

Design and interpretation issues

9.1 Introduction

In the first part of this chapter we consider problems that have to be dealt with by the analyst as a result of shortcomings in either the collection of the data or in the way that the study has been conducted.

One perennial problem encountered in many types of studies, no matter how hard investigators try to prevent it, is that of missing data. Values in a dataset can be missing for all sorts of uncontrollable reasons: malfunctioning equipment, incorrect data entry, illness on the day of data collection, unwillingness of a respondent to answer some questions, and so on. In the first section below we give a general categorization of mechanisms that lead to missing values, and discuss the implications of this categorization for handling missing values in the context of ROC analysis.

A second major problem area arises when the individuals in a study are in some way unrepresentative of the population about which information is being sought, which can then lead to bias in the conclusions drawn from the data. Many forms of bias are as relevant to ROC studies as to other studies, and these are briefly reviewed at the start of the second section below. However, there is one form that is specific to the ROC case, especially in medical diagnosis contexts, and this form is discussed more extensively in the second section. It occurs when the true state of the patient can only be determined by a gold standard test that is either very expensive, or very invasive, or has attendant dangers regarding the patient's safety. Such a gold standard test is generally only applied very sparingly, and if the group to which it has been applied is very selective but the results are interpreted as if it had been randomly applied, then the interpretation is biased. This type of bias is known as verification, or work-up, bias. We discuss its genesis, and describe the methods that have been proposed for correcting for it.

In the next part of the chapter we turn to an interpretational issue, namely how to find from a ROC curve the optimum threshold at which to operate the associated classifier. This is an important question when a practical diagnostic procedure actually needs to be conducted; so an optimum performance is the objective. This problem was briefly addressed in Section 3.5.3 assuming that costs of misallocating individuals could be quantified and assigned, but in many practical situations (in particular medical ones) such assignment is at best questionable. Possible solutions that do not need this assignment are discussed in the third section below.

In the final part of the chapter we consider a particular subject area, namely medical imaging, in which many applications of ROC analysis may be found, and we summarize the design and interpretation issues that arise in this area.

9.2 Missing values

9.2.1 Missing value mechanisms

Gaps can arise in data matrices for a variety of reasons. The problem has long been recognized, and many methods have been devised for overcoming it. In the early days a "one size fits all" approach was adopted, in the hope that a single method (generally fairly crude) would suffice whatever the reason for the missing data values. However, as research continued into more sophisticated methodology, it became evident that different reasons for "missingness" led to different inferential problems, and if a particular method was appropriate for one reason it might no longer be appropriate for another. This led in turn to a categorization of such reasons, or "mechanisms," that produce missing values, and the following three mechanisms are now widely recognized (Little and Rubin, 2002).

9.2. MISSING VALUES

1. Missing completely at random (MCAR)
   This is when the probability of an observation being missing does not depend on any observed or unobserved measurements. An alternative way of viewing it is to say that the mechanism is MCAR if missing values are randomly distributed across all observations. An example would be if all the samples being delivered for chemical analysis on one day, from a variety of sources, were accidentally destroyed and hence the results of these analyses were unavailable. There is no reason to suppose that any systematic pattern exists among the samples that are missing. However, one has to be careful: a mechanism which seems to be MCAR may turn out not to be. For example, suppose these chemical analyses were typically conducted at several different chemical laboratories, and results were then sent on to a central record office. Missing results might be taken as MCAR but could in fact have a systematic cause, such as the delivery procedure from the different laboratories. So, for example, if the samples destroyed had all come from the same laboratory then MCAR would not be an appropriate assumption.

2. Missing at random (MAR)
   This category takes care of the last example in the paragraph above and can be expressed in three equivalent ways: (i) the probability of an observation being missing depends only on the observed data; (ii) given the observed data the probability of an observation being missing does not depend on unobserved data; (iii) the missing values are randomly distributed within one or more subsamples. In the above example, the probability of sample results being lost might differ between laboratories but lost results are randomly distributed within each laboratory, so the missing values are MAR rather than MCAR. Another example would be if in a survey there were nonresponses to a particular question and these nonresponses varied between different ethnic groups, but were random within each group.

3. Missing not at random (MNAR)
   This is the category for all cases that are neither MCAR nor MAR, and is appropriate whenever the probability of an observation being missing depends on an unseen value (either the missing value itself, or some other unmeasured variable). In this case,

missing values are not randomly distributed across observations but form some type of selected group. Moreover, the probability that an observation will be missing cannot be predicted from the other measured variables. Examples occur when collecting information about income, as people may have a variety of unspecified reasons for not revealing their income; or in radiography where missing measurements may have arisen because the doctor made a judgement from personal knowledge of the patient's medical history that a particular test would not be worth conducting.

Distinguishing MCAR and MAR can be done quite easily, either using simple t-tests or from readily available statistical software. For example, dividing individuals into those with and those without missing values and conducting t-tests of mean differences between the groups on the various key variables in the study will confirm MCAR if the tests are predominantly nonsignificant, while statistical packages such as SPSS have procedures that test for MAR. Unfortunately, it is not possible to distinguish MNAR from either of the other two categories, using the data at hand.

MCAR and MAR are often termed *ignorable* missing value mechanisms, on the grounds that methods for dealing with the missing values do not have to take into account the mechanism that generated them. By contrast, MNAR is a *nonignorable* mechanism, and to obtain valid inferences a joint model of the data and the missingness mechanism has to be used. Unfortunately, in the MNAR setting, an appropriate model may be very difficult to find or formulate. The way forward would be to conduct sensitivity analyses, exploring the differences in inferences reached under different models and different assumptions about the mechanism, but the extent to which this is possible will of course depend on the available time and budget.

9.2.2 General missing value methodology

Much work has been done over the years, and many different methods have been proposed, for dealing with missing values when analyzing a data set. We here provide just a brief overview, so readers wishing to follow up any details will need to consult a specialized text (e.g., Allison, 2001, for a concise account or Little and Rubin, 2002, for a comprehensive treatment). Broadly speaking, there are three general strategies for handling missing values: deletion of some observed values;

9.2. MISSING VALUES

imputation of estimates for the missing values; or maximum likelihood methods that utilize all the observed values and only these values. The first two strategies are appropriate whenever the chosen method of analysis requires either a complete data matrix or its respective summary statistics (such as mean vectors and covariance matrices), while the third strategy requires an underlying probability model for which population parameters are to be estimated. We now briefly consider each strategy, highlighting any strengths and weaknesses and relating it to the above missing value mechanisms.

Deletion of observed values

If the chosen analysis requires a complete data matrix then the simplest approach is to omit those cases that have *any* values missing, and to run the analysis on what remains. This approach is termed *casewise* (or *listwise*) deletion. Its benefit is that if the data are MCAR then the analysis and parameter estimates are unbiased. However, it may lead to a substantial decrease in the sample size, so that even if the data are MCAR there will be a loss of power. Moreover, if the data are *not* MCAR then the results are likely to be biased, and this bias may be serious if the missing values occur in some very systematic way.

If many variables have been measured and there is a moderate overall proportion of missing values, but only one or two at most are missing for any single case, then casewise deletion will result in many observed values being omitted. Some analysts do not like this, and look for ways in which as many of the observed values as possible can be used. One way of achieving this occurs when the analysis requires only estimates of mean vectors and covariance or correlation matrices, rather than presence of a complete data matrix. In this case each element of the mean vectors can be obtained by averaging all values that are present on the relevant variable , and each element of the matrices can be obtained separately from all values present for the relevant pair of variables. This approach is usually termed *pairwise* deletion, but is not to be recommended in general. The main problem is that every element of both the vector and the matrix is estimated from different sets of data, which implies that each will have a different standard error so that inferences from the analysis will not be reliable. Moreover, matrices formed in this way are frequently not positive definite, in which case the analysis may fail entirely.

Imputation of missing values

Instead of deleting observed values, many analysts prefer to replace missing values by some plausible estimates and thereby to "complete" the data matrix for analysis. The oldest suggestion in this respect is just to replace any missing value by the (overall) mean of the relevant variable. This is often termed *unconditional* imputation. However, this approach imputes the same value for a given variable in different cases, so if the data are not MCAR then such a process will make any inferences biased. A better way of estimating any single missing value would be as the fitted value in the multiple regression of its associated variable on all the other variables. That way the imputed value is *conditional* on the information we have for the other variables on that case; so this approach will be valid if the data are not just MCAR but also MAR. However, if one of the variables in the data set is a dependent variable while the others are explanatory ones, then missing values will generally be restricted to the latter set. In this case one common approach is to create an additional variable for each variable containing missing values and to code this new variable 0 or 1 according as values are present or absent on its partner. Reversing the signs of the estimated regression coefficients of the new variables produces the imputed values. This is often termed the *missing-indicator* method.

A problem with all of these methods is that the imputed values are perfectly predictable, either from the single variable or from a combination of all the variables, so no new information has been added but sample sizes have in effect been increased because more "values" are now present in the data set. Consequently, standard errors are artificially reduced (*attenuated*) and inferences may thereby be skewed. There are several ways in which the sizes of standard errors can be preserved. One is *stochastic imputation*, in which a random value is added to each imputation. This random value can be obtained in various ways, most usually either by generating a random observation from a (Normal) distribution having mean zero and variance equal to the variance of the observed values for the relevant variable, or by sampling with replacement from the residuals in the regression used to fit the imputed value. Another possibility is *multiple imputation* in which a number k of separate imputations are made for each missing value, to represent the uncertainty about the correct value to impute. This is done either by effecting k independent stochastic imputations or by using the structure of the data in some appropriate way to generate the

9.2. MISSING VALUES

imputations. This leads to k completed data matrices, each of which is analyzed separately and then the results of the analyses are combined to produce the final inferences. Stochastic imputation reduces the attenuation problem but may not totally eliminate it; multiple imputation provides the best remedy and has therefore become popular (see Schafer, 1999, for further details).

Maximum likelihood

This strategy is appropriate whenever a probability model underlies the data, and parameters of the model are to be estimated. For example, the rows of the data matrix may be assumed to come from a Normal distribution with mean vector $\boldsymbol{\mu}$ and dispersion matrix $\boldsymbol{\Sigma}$. The likelihood of the sample is just the product of the Normal probability densities at the data values, and maximum likelihood estimates (MLEs) of the parameters are the values that maximize this quantity. If there are missing values in the data, then the MLEs can be obtained using just the observed values in a series of iterations of the so-called EM algorithm. Each iteration has two components: maximum likelihood estimation of parameters using currently available data, then prediction of the missing values using the given probability model. The starting point is usually estimation after casewise deletion, and the process is stopped when two successive sets of estimates differ by less than some preset tolerance. Note that prediction of missing values in a probability model is done most naturally through conditional expectations, and in the case of the Normal model above the conditional expectation of any variable is its linear regression on the other variables. So the MLEs in this case are obtained by iterating the regression imputation process of the previous section.

9.2.3 Some implications for ROC studies

In a sense, missing values cause most problems at the pre-ROC stage, as they have to be dealt with in order to produce the classifier from which the ROC curve is obtained. The preceding discussion of missing values has been included mainly to provide a foundation for the section on verification bias that follows, but a few articles have recently appeared that connect missing values and ROC analysis in various ways so it is appropriate to mention them here.

Van der Heiden *et al.* (2006) have in fact investigated the pre-

ROC problem in clinical diagnosis. They applied five different methods for handling missing values (casewise deletion, missing-indicator, single unconditional and conditional mean imputations, and multiple imputation) to a data set of 398 subjects in a prospective study on the diagnosis of pulmonary embolism. For each method they fitted a diagnostic prediction model using multivariable logistic regression analysis, selected the best variables for prediction, and compared the fitted models by means of ROC curves and AUC values. They found that the chosen predictors for the casewise deletion and missing-indicator methods were very different from those chosen for the other methods, and the chosen predictors for all three imputation methods were very consistent. The models based on multiple imputation did not differ much from those derived after either of the single imputation methods. The title of the article entirely describes their conclusion, namely that imputation is superior to casewise deletion and missing-indicator analysis for clinical diagnosis.

Lin and Haug (2008) conducted a similar investigation, but this time comparing four procedures (no action, single mean imputation, missing-indicator method, and use of an extra categorical state "missing") on the development of a Bayesian Network expert system for clinical diagnosis. The experiment was carried out on a large data set using the area under a ROC curve to evaluate performance of methods. They found that the missing-indicator method and the extra categorical state representing "missing" performed better than either no action or single mean imputation, so recommended that expert systems should incorporate explicit models of missingness for best results.

The final article, by Spritzler *et al.* (2008), has the least connection with ROC but merits a mention because it concerns adjustments necessary to two-sample tests on areas under curves in the presence of missing values. The curves here are not ROC curves, but rather profiles over time of groups of subjects submitted to two different treatments or regimes. For example, in a clinical trial patients suffering from a serious disease are randomized to one of two competing treatments and the response (e.g., blood pressure) is measured several times over a fixed period. The average measurements for each treatment group are plotted over time points, and the two resulting profiles of points constitute the curves to be compared. Areas under the two curves are computed by the trapezoidal method (Section 3.5.1); a significant difference between the two computed areas is indicative of a difference in effect between the two treatments. Spritzler *et al.* (2008) give the necessary calculations using all the available data to obtain a test based on estimators that are unbiased when the data are MCAR. They then go on to develop a semi-parametric test that has good properties when the missing data mechanism is relaxed to MAR, but this test is only appropriate when the missingness is monotonic (i.e., once a variable is missing then it is always missing subsequently, as happens in clinical trials whenever patients drop out of the trial).

9.3 Bias in ROC studies

9.3.1 General

In all practical instances of ROC analysis the ROC curve has been constructed using a classifier built from a *sample* of individuals, while any conclusions that are drawn from the analysis are intended to apply to the whole *population* from which the sample has come. *Bias* will occur if these conclusions deviate from the truth by more than the chance vagaries of sampling, and will reflect some systematic error or shortcoming in the whole process. Such systematic error can be introduced either because of some technical deficiency in the construction of the classifier, or because of some failing of the sampling process. The former cause lies outside the scope of the present book, so readers interested in following up work on bias considerations in the construction of classifiers are referred to specialist texts such as that by McLachlan (1992). We will concentrate here on the latter cause. Moreover, since by far the most important application area for considerations of bias is medical diagnosis, we will present most of our results and examples in terms of this area.

To avoid the risk of introducing systematic deviation between features of the sample and those of the population, the researcher should ensure that all members of the target population are potentially able to be included in the sample, and that the selection mechanism does not prevent any recognizable subgroups from being included. In many areas of statistics these matters can be ensured by listing populations in some way, and using a random device (e.g., pseudo-random numbers) for picking items from the list. Various sophisticated refinements of these basic operations have been developed over the years to deal with more complex populations, and are described in text books on sampling methods (e.g., Cochran, 1977). However, there are some specific problems that the researcher needs to look out for. *Selection* bias will occur when the composition of the sample is influenced in some way by external factors, and *spectrum* bias when the sample does not include the complete spectrum of patient characteristics. Often there is some connection between these two types of bias. For example, suppose that a classifier is being developed to test for the presence of condition A, and the subjects for the study are being recruited by referral from family doctors. It is not usually possible to determine referral policies of individual doctors, so if one doctor tends to refer all patients who exhibit condition B but only some who do not exhibit B, then there will be an excess of the former in the sample. This is certainly an example of selection bias, while if only patients exhibiting condition B are referred, then that is likely to lead to spectrum bias.

These two types of bias can only be removed by appropriately designing the study, and adjusting results post-hoc is virtually impossible as the true population characteristics are generally unknown. However, there is another type of bias common in ROC studies for which much work has been done on such post-hoc adjustments, so we now concentrate on this type.

9.3.2 Verification bias

Suppose that a new classifier with continuous classification score S has been proposed for screening individuals for some condition A, say. As usual, we will assume that an individual is allocated to population P (i.e., condition A is present) if his or her score s exceeds some threshold t, and to population N (i.e., condition A is absent) otherwise. The accuracy of this classifier can be assessed (ideally) by comparing it with a perfect gold standard (GS) test that determines presence or absence of condition A with certainty. Following the procedures outlined in previous chapters, a random sample of n individuals who are to be tested is subjected to both classifiers, the classification provided by the GS test are taken as the true grouping, and the results of the new classifier are related to the true grouping via a ROC curve. Accuracy assessments can then be made using either AUC or some of the other related quantities.

However, the GS might be either too expensive or too invasive for everyday use. While the new classifier can be applied to all n sample individuals, it may not be feasible to subject them all to the GS. If only a subsample can be chosen for the GS verification, then ethical

9.3. BIAS IN ROC STUDIES

considerations would dictate biasing this subsample (the "verification" sample) towards those individuals whose associated signs, symptoms, or medical history suggest presence of condition A. Unfortunately this induces a special form of selection bias, known variously as "verification bias" (Begg and Greenes, 1983) or "work-up bias" (Ransohoff and Feinstein, 1978), and incorrect conclusions can result from a ROC analysis that assumes the individuals to have been randomly sampled. Note that this situation can be formally treated as a "missing value" one, since the true group status is missing for those individuals that have not been selected for the GS.

In such circumstances, either bias-corrected methods or ones robust against the presence of verification bias are needed. While many published methods do exist, nearly all of them are limited to either dichotomous or ordinal classification scores S (see Rotnitzky et al., 2006, for an extensive list) and only two are currently available for the case when S is continuous. One of them makes the assumption of an ignorable (MAR) missing value mechanism, while the other is concerned with a nonignorable (MNAR) mechanism. We now briefly consider these two approaches.

Alonzo and Pepe (2005) focus on the simple empirical ROC curve in which estimates of $tp(t)$ and $fp(t)$ are plotted for a range of threshold values (Section 3.3.1), with AUC being obtained by the trapezoidal method (Section 3.5.1). Suppose that n_v individuals have been subjected to the GS so are in the verification sample, and $n_v(A)$ of these have condition A while $n_v(\bar{A})$ do not have condition A. Thus $n_o = n - n_v$ individuals are not included in the verification sample. Let $C(A, v)$ and $C(\bar{A}, v)$ denote the sets of individuals in the verification sample that respectively have and don't have condition A, $C(v)$ the set of all individuals in the verification sample, and $C(o)$ the set of individuals not in the verification sample. The standard empirical ROC curve is derived by applying the usual estimators to just the individuals in the verification sample. This corresponds to casewise deletion of the individuals in $C(o)$, and can be termed *complete case* (CC) analysis. The required estimators are thus given by

$$\widehat{tp}_{CC}(t) = \frac{1}{n_v(A)} \sum_{i \in C(A,v)} I(s_i \geq t), \quad \widehat{fp}_{CC}(t) = \frac{1}{n_v(\bar{A})} \sum_{i \in C(\bar{A},v)} I(s_i \geq t)$$

where, as usual, $I(\mathcal{S}) = 1$ if $\mathcal{S}$ is true and 0 otherwise.

Of course, casewise deletion only gives unbiased estimators if the missing values are MCAR; so these CC estimators will only be useful if the verification sample is a true random sample of all n original individuals. As this is usually very unlikely, Alonzo and Pepe (2005) derive a number of bias-corrected estimators under the less restrictive MAR assumption. They do this by making two connections. First, by writing $tp(t)$ as $\Pr(S \geq t, A)/\Pr(A)$ and $fp(t)$ as $\Pr(S \geq t, \bar{A})/\Pr(\bar{A})$ they note that $tp(t)$ and $fp(t)$ can be thought of as ratios of prevalence probabilities. Second, a study with verification-biased sampling can be thought of as one with a double-sampling or two-phase design: selection and measurement of n individuals for the first phase of the study, followed by selection of the sample for verification at the second phase. The probability of selection for the second sample is dependent only on information gathered at the first phase, so the MAR assumption holds. This allows methods for estimating disease prevalences in two-phase designs (Gao et al., 2000; Alonzo et al., 2003) to be adapted to bias-corrected estimation of $tp(t)$ and $fp(t)$.

Alonzo and Pepe (2005) propose the following bias-corrected estimates developed from these connections.

(i) Full imputation (FI).
Here, the probability $\rho = \Pr(A|S)$ is estimated by fitting a logistic regression to the verification sample, from which we obtain $\widehat{\Pr}(A) = \frac{1}{n}\sum_{i=1}^{n} \hat{\rho}_i$ to estimate the denominator in $tp(t)$ above. For the numerator we have

$$\Pr(S \geq t, A) = \int_{-\infty}^{\infty} I(S \geq t) \Pr(A|S = s) \Pr(S = s) ds,$$

so that an empirical estimator is given by

$$\widehat{\Pr}(S \geq t, A) = \frac{1}{n} \sum_{i=1}^{n} I(s_i \geq t) \widehat{\Pr}(A|s_i).$$

Hence the FI estimator for tp is

$$\widehat{tp}_{FI}(t) = \frac{\sum_{i=1}^{n} I(s_i \geq t) \hat{\rho}_i}{\sum_{i=1}^{n} \hat{\rho}_i},$$

with the obvious analogous estimator for fp.

(ii) Mean score imputation (MSI).
Here, a logistic regression is again fitted to the verification sample to estimate $\rho = \Pr(A|S)$. However, this estimate is only imputed

9.3. BIAS IN ROC STUDIES

for those individuals not in the verification sample, while the observed incidences are used for individuals in the verification sample. Thus

$$\widehat{tp}_{MSI}(t) = \frac{\sum_{i \in C(A,v)} I(s_i \geq t) + \sum_{i \in C(o)} I(s_i \geq t)\hat{\rho}_i}{n_v(A) + \sum_{i \in C(o)} \hat{\rho}_i}$$

and analogously for fp.

(iii) Inverse probability weighting (IPW).

In some situations, the probability π_i of an individual with $S = s_i$ being included in the verification sample may either be known or may be estimable from the study design. In this situation a complete case approach can be used, but weighting each observation in the verification sample by the inverse of the probability of that individual's selection. This yields the estimator

$$\widehat{tp}_{IPW}(t) = \frac{\sum_{i \in C(A,v)} I(s_i \geq t)/\hat{\pi}_i}{\sum_{i \in C(A,v)} (1/\hat{\pi}_i)}$$

and analogously for fp.

(iv) Semiparametric efficient (SPE).

Gao et al. (2000) and Alonzo et al. (2003) used parametric models for ρ and π, and a nonparametric approach for the data, to derive semiparametric estimators of disease prevalence in two-phase studies. This approach leads to the estimator $\widehat{tp}_{SPE}(t)$ given by

$$\frac{\sum_{i \in C(A,v)} I(s_i \geq t)/\hat{\pi}_i - \sum_{i \in C(v)} I(s_i \geq t)\hat{\rho}_i/\hat{\pi}_i + \sum_{i=1}^{n} I(s_i \geq t)\hat{\rho}_i}{\sum_{i \in C(A,v)} (1/\hat{\pi}_i) - \sum_{i \in C(v)} \hat{\rho}_i/\hat{\pi}_i + \sum_{i=1}^{n} \hat{\rho}_i}$$

and an analogous expression for the estimator of fp.

Alonzo and Pepe (2005) give some asymptotic theory, from which consistency of estimators can be deduced and confidence intervals can be obtained. They also compare the various estimators with respect to bias, relative efficiency, and robustness to model misspecification, and give some practical examples of their implementation.

Rotnitzky et al. (2006) directly tackle the estimation of AUC in the presence of verification bias, but under the wider assumption of a MNAR mechanism rather than MAR for the cases that have not been selected for verification. They also assume that in the absence of missing values AUC would be estimated using the Wilcoxon rank-sum statistic, which is an alternative way of calculating the Mann-Whitney U-statistic (Section 3.5.1).

Since an MNAR mechanism is assumed, for valid inferences a model linking data and missingness mechanism must be specified (see Section 9.2.1 above). Let us denote the true condition state of a patient by D (values 1/0 for presence/absence of condition A); that patient's membership of the verification sample by R (values 1/0 for presence in or absence from the sample); and classification score by S and covariates by $\mathbf{Z}$, as usual. Then Rotnitzky *et al.* (2006) model the logarithm of the ratio of probabilities of a patient being included or excluded from the verification sample by a linear combination of an unknown function $h(S, \mathbf{Z})$ and D times a user-specified function $q(S, \mathbf{Z})$. This is equivalent to saying that the conditional probability of the ith subject being in the verification sample, given that subject's measured values S_i, $\mathbf{Z}_i$ and condition status D_i, is

$$\pi_i = [1 + \exp\{h(S_i, \mathbf{Z}_i) + q(S_i, \mathbf{Z}_i)D_i\}]^{-1}.$$

Rotnitzky *et al.* (2006) then propose a sensitivity analysis strategy, under which the analyst investigates the AUC estimates across a range of plausible selection bias functions $q(S, \mathbf{Z})$. However, problems arise when both S and $\mathbf{Z}$ are continuous because of the curse of dimensionality, and to make progress a parametric model has to be assumed either for $q(S, \mathbf{Z})$ or for the log ratio of probabilities of D values in the verification sample. Rotnitzky *et al.* (2006) propose suitable estimators of the parameters of these models, and define an imputation function involving S, $\mathbf{Z}$ and both of these estimated models. An estimate of AUC is then obtained by applying this function to those subjects not in the verification sample, imputing the resulting value in place of the missing condition status, and applying the usual Mann-Whitney U-statistic to the resulting completed set of data. They prove that the expectation of this imputed value is equal to the expectation of the corresponding condition status providing that *either* of the assumed parametric models has been correctly specified, and hence term their estimator the *doubly robust* estimator of AUC.

In the most recent contribution concerning verification bias, Chi and Zhou (2008) extend the methodology to encompass three-group ROC surfaces. Their study is restricted to ordinal classification scores,

9.3. BIAS IN ROC STUDIES

so strictly speaking is outside the remit of this book, but the article is worth a mention as it is the first one to consider verification bias outside the usual two-group situation. The authors assume an ignorable (MAR) missing value mechanism, develop an estimator for the volume under the ROC surface in the presence of verification bias, and use Fisher's Information and jackknifing to derive standard errors. They also propose necessary adjustments to allow for the correlation induced when the interest focusses on the difference between two volumes, and conduct extensive simulation studies to investigate the accuracies of the various estimators.

9.3.3 Reject inference

A particular form of selection bias arises when screening instruments are constructed based on cases which have themselves been chosen by previous application of another screening instrument. This is a ubiquitous problem in areas such as consumer credit scoring. Here, applicants are accepted (e.g., for a loan, credit card, or other financial product) if they exceed the threshold on some "creditworthiness" score (that is, if they are classified into the "good risk" class by a classifier). Some of these customers will subsequently default, though one might hope that most will not. Then, using the observed characteristics of the accepted customers, as well as their known eventual outcome (default or not), one will seek to construct a new, updated, creditworthiness score, to apply to future applicants. The trouble is, of course, that while one intends to apply the new scorecard to *all* future applicants, the data on which it will be based are the subset of applicants who were previously thought to be good using the original classifier. If the original classifier had any predictive ability in distinguishing between the default/nondefault classes, the resulting data set is likely to be heavily distorted in favor of good risks: it is not a representative sample from the population to which it is to be applied.

This can have serious consequences. For example, suppose that a particular binary variable used in the original classifier was highly predictive—with, for example, a value of 0 indicating very high risk. It is likely that very few people with a value of 0 on this variable will have been accepted. That means that, when a new classifier is constructed using the screened applicants, very few of them will take a value of 0 on this variable. That, in turn, will mean that it is unlikely to be identified as an important contributor in distinguishing between defaulters and nondefaulters—so it will not appear as a predictor variable in the new classifier. In short, the selection mechanism has led to the dropping of a highly predictive variable.

This problem is widely recognized in the consumer banking industry, and a variety of methods have been explored in attempts to overcome it. Because many of them are based on attempts to estimate the unknown outcome class of those applicants who were rejected by the initial classifier, the general area has become known as *reject inference*. Since, by definition, the outcome class of rejected applicants cannot be known, to tackle the problem it is necessary to obtain extra information from somewhere else. This may come in various forms, including assumptions about the shapes of underlying distributions, accepting some customers who were classified as likely defaulters by the original classifier, and following up those rejected applicants who are successful in obtaining financial products from other lenders to obtain a proxy outcome indicator.

Although most attention in the consumer banking industry has focussed on tackling the sample distortion when constructing new classifiers, the problem also impacts the evaluation of classifiers. In particular, since, by definition, the original classifier has rejected any applicant with a score less than some threshold value, the distributions of the defaulters and nondefaulters amongst the accepted applicants will be less well separated (e.g., using measures based on the ROC curves) than would be the corresponding score distributions of the overall population. Since the distributions of scores on another classifier not used for the selection will not be truncated in this way, a measure of separability for this other classifier will not be subject to the same downward bias. The consequence is that comparisons of separability measures between the original classifier (used for the initial screening) and the new one may be biased in favor of the original classifier.

9.4 Choice of optimum threshold

Any ROC curve, or a summary value such as AUC and PAUC computed from it, only shows the overall worth of a classifier across all possible choices of classification threshold. However, if the classifier is to be applied in practice to a group of individuals requiring diagnosis, say, then a unique threshold must be chosen so that a single classification is produced for each individual. If it is possible to quan-

9.4. CHOICE OF OPTIMUM THRESHOLD

tify the costs incurred in misclassifying individuals, then the method of Section 3.5.3 can be used to find the threshold that minimizes the expected cost due to misallocation of individuals. However, quantification of costs can be arbitrary, unreliable, or even impossible to provide, and in the absence of such costs investigators often assume that they are all equal and adopt some pragmatic procedure for optimum threshold determination. One seemingly reasonable approach is to locate the point on the ROC curve nearest the top left-hand corner of the plot, and to use its corresponding threshold. However, properties of this procedure have not been fully determined and, moreover, Perkins and Schisterman (2006) have cautioned against it on the grounds that it can introduce an increased rate of misclassification.

A systematic approach requires a meaningful criterion to be formulated in such a way that optimizing the criterion leads to determination of a unique threshold. The criterion that has received most attention in the literature is the Youden Index $YI = \max(tp - fp) = \max(tp + tn - 1)$, and this is the criterion used by Fluss et al. (2005) and Le (2006) in order to determine an optimum threshold for a single classifier. Indeed, to establish that it provides a meaningful criterion, Le notes two important characteristics of YI, namely: (i) a process can only qualify as a genuine diagnostic test if it selects diseased individuals with higher probability than by chance, and this is so if and only if $YI > 0$ for the process; and (ii) the precision of the empirical estimate of disease prevalence depends only on the size of YI and not on any other function of sensitivity or specificity.

Let us return to the notation that we established in Section 2.2.4 and used throughout subsequent chapters, that the density and distribution functions of the continuous classification score S are given by $f(x), F(x)$ in population N and $g(x), G(x)$ in population P. Then Fluss et al (2005) note that $YI = \max(tp - fp) = \max(tp + tn - 1) = \max_t[F(t) - G(t)]$, so that YI can be estimated using any of the estimators for F, G described in Chapter 3. The optimum threshold t^* is the value of t that maximizes the estimated $F(t) - G(t)$; so different estimators will in general yield different YI and t^* values. Fluss et al. (2005) consider four possible methods, the first two being parametric and the remaining two nonparametric.

(i) The normal method.
   Here it is assumed that $S \sim N(\mu_N, \sigma_N^2)$ in population N and $S \sim N(\mu_P, \sigma_P^2)$ in population P, whence YI is the maximum over

t of $\Phi\left(\frac{t-\mu_N}{\sigma_N}\right) - \Phi\left(\frac{t-\mu_P}{\sigma_P}\right)$. Computing the first derivative of this expression, setting it to zero and solving the resulting quadratic equation, gives the optimum threshold t^* as

$$\frac{(\mu_P \sigma_N^2 - \mu_N \sigma_P^2) - \sigma_N \sigma_P \sqrt{(\mu_N - \mu_P)^2 + (\sigma_N^2 - \sigma_P^2)\log(\sigma_N^2/\sigma_P^2)}}{(\sigma_N^2 - \sigma_P^2)}.$$

If equal variances are assumed, i.e., $\sigma_N^2 = \sigma_P^2 = \sigma^2$, then the result is much simpler: $t^* = (\mu_N + \mu_P)/2$.

YI and t^* are estimated by substituting sample means and variances for the corresponding population values in these expressions. However, the assumption of normality may not be realistic in many situations.

(ii) The transformed normal method.

A less restrictive approach is to assume that there exists some monotonic transformation U such that $U(S)$ is normal in populations N and P. Various authors have recommended using the data to fit a Box-Cox power transformation of the form

$$u = \begin{cases} (s^\lambda - 1)/\lambda & \lambda \neq 0, \\ \log(s) & \lambda = 0. \end{cases}$$

Assuming that the transformed values have normal distributions, λ can be estimated by maximum likelihood, the data can then be transformed, and normal method (i) can be applied. This approach should be satisfactory in many practical cases, but if it is evident from the form of the data that an adequate transformation to normality is unlikely then a nonparametric method must be employed.

(iii) Empirical method.

Here the cdfs F and G can be estimated using the empirical cdfs of the two samples. The estimate of $[F(t) - G(t)]$ can thus be obtained directly at each value in the two samples, whence an estimate of t^* is given by the sample value which maximizes the estimate of $[F(t) - G(t)]$. Fluss et al. (2005) suggest that a slightly preferable estimate might be half-way between this value and the one just above it when values from both samples are merged and arranged in ascending order. However, while this is an easily implemented method it may be overly subject to sampling vagaries, and some form of smoothing might be preferred.

9.4. CHOICE OF OPTIMUM THRESHOLD

(iv) Kernel method.

Here the cdfs F, G are estimated using the kernel density method described in Section 3.3.3, and this again leads directly to an estimate of $[F(t) - G(t)]$ at each value in the two samples. However, iterative numerical methods now need to be used to find the estimates of YI and t^*.

Fluss et al. (2005) apply all four methods to a practical example, namely diagnosis of Duchenne muscular dystrophy. This example demonstrates that there is considerable variability in the estimates of t^*, while estimates of YI are rather more consistent. They then report extensive simulation studies which suggest that the method most commonly used in practice, namely the empirical cdf method, actually has the worst performance. In terms of estimating YI the Kernel method is generally best, although the normal or transformed normal methods are good whenever the assumptions are justified. Estimates of t^* are much more variable than those of YI.

Rather than starting from assumptions about the classification statistic S, Le (2006) starts from the ROC curve itself and supposes that it is fitted by a smooth model of the form $y = R(x)$. Then $tp - fp = R(x) - x$, so the strategy is first to find the value x_{opt} which maximizes $R(x) - x$. Le suggests modeling the ROC curve by the proportional hazards model $R(x) = x^\theta$ for $0 \leq x \leq 1$, whence

$$\frac{d}{dx}[R(x) - x] = \theta x^{\theta-1} - 1.$$

Setting the derivative to zero and solving the resulting equation gives $x_{opt} = \frac{1}{\theta^{1/(\theta-1)}}$. Moreover, AUC $= \int_0^1 x^\theta dx = \frac{1}{\theta+1}$, so that $\theta = \frac{1-\text{AUC}}{\text{AUC}}$ and x_{opt} can be easily evaluated in terms of AUC.

These quantities suggest a simple process for estimation in practice. Given samples of classification scores from each of the two populations N and P, AUC is estimated using the Mann-Whitney U-statistic (Section 3.5.1). This provides estimates of θ and hence x_{opt} by substituting in the above expressions. But x_{opt} is one minus the cdf of population N at the optimal threshold t_{opt}, so this optimal threshold can thus be deduced as the cut-point which produces the proportion $1 - x_{opt}$ of the empirical distribution function of the sample from the population N. Le (2006) gives several numerical examples of this procedure, and concludes his paper with a discussion of testing the fit of the models.

9.5 Medical imaging

One of the first areas outside psychology to make use of ROC analysis was that of diagnostic radiology (Lusted, 1968; Goodenough *et al.*, 1974), and this is where many applications of such analysis are now encountered. Diagnosis is typically made from interpretation of a medical image by a trained observer/expert; in radiology, patients are usually referred to as "cases" and observers as "readers." Very often the outcome is either a nominal or ordinal value (e.g., diseased/not diseased or normal/mildly affected/seriously affected), but it might equally be a continuous value such as a percentage, a confidence score, a probability assessment, or a summary accuracy measure from a fitted ROC curve such as AUC or PAUC, so will come within the scope of this book. We first consider briefly some design issues and then turn to interpretation and analysis questions, specifically for multiple reader studies.

9.5.1 Design

Metz (1989) sketches out some of the practical issues that arise in ROC studies of radiological imaging. He first discusses the pros and cons of random sampling versus constrained selection for obtaining the cases to be used in the ROC study: random sampling ensures unbiasedness of estimates but at the expense of large variance in small samples, while stratified sampling provides greater precision of estimates but at the possible expense of bias if incorrect information is used about the strata prevalences. He then notes that studies attempting to measure absolute detectabilities achieved by several classifiers are much more challenging than ones seeking merely to rank the classifiers on the basis of their detectabilities. The former study requires extreme care to ensure that all aspects of the diagnostic settings of interest are represented accurately and all potential biases are avoided, while the latter study is less demanding and allows the individual designing it much greater freedom. The final design issue he considers is that of reading-order effects in studies where the readers are required to assess two or more equivalent images of a particular case. In order to eliminate "carry-over" effects, whereby the reader uses knowledge gained from the first image when assessing the second one, some randomizing of the order in which they are assessed is necessary and Metz gives several possible plans.

Begg (1989) focusses on comparative studies and discusses the pros and cons of matched studies (where several classifiers are applied to

9.5. MEDICAL IMAGING

each case) versus randomization (where each classifier is applied to a separate sample of cases). Readers will perhaps be already familiar with statistical considerations underlying the differences between matched pairs and separate samples for hypothesis tests about the difference in means between two populations, and similar considerations come into play in ROC studies. Where patient outcome is an end point, as in cancer screening studies, then randomization is the relevant method. For accuracy studies, however, matching provides much greater efficiency and internal control, but there is serious potential risk of bias in subjective assessment of images if information from the matched test is available. Blinding of tests and randomizing their sequence are again necessary precautions to offset this problem.

Gur *et al.* (1990) focus on the selection of negative controls in ROC studies, using some theoretical considerations and experimental data to demonstrate the significance of the issue. Their data come from diagnoses made by observers from medical images, and they comment that the results will be generally relevant to situations where a diagnosis of "no abnormality" is likely to be made with a high degree of confidence in a large fraction of the control images presented for study. Because it focusses on binary outcomes, however, this article is only of tangential relevance to the present volume.

Since 1990, much emphasis has been placed on "multiple reader, multiple case (MRMC)" studies. The inherent variability in readers' accuracies was appreciated from the start; so for reliable inferences to be possible it was felt necessary to involve several readers as well as several cases in each study (see also Obuchowski, 2004). Between 4 and 15 readers are commonly used, and studies are generally designed as fully factorial ones where the same cases undergo all the diagnostic tests under study and the same readers assess the results from all of these tests. If all cases have undergone all tests and all readers have assessed all results, then the design is a "paired-patient, paired-reader" one, but it is possible to have any one of four designs by attaching either of paired/unpaired to either of patient/reader. Whichever design is chosen, there will be distinct sources of error arising from variability between cases (because of disease characteristics, say) and variability between readers (because of differences in skills, training, or experience) as well as potential correlations between readers' interpretations and between different test results. So the bulk of the effort put into MRMC studies has been concentrated on analysis and interpretation, particularly on allowing for reader variability.

9.5.2 Analysis and interpretation

Various methods of analyzing data from MRMC studies have been proposed. A statistical summary of each, and comparison between them using previously published datasets, has been given by Obuchowski *et al.* (2004). We here give just a brief overview; technical details may be found in the cited references.

The first methods to be proposed were parametric, centered on analysis of variance. Presence of all tests on each reader and case suggests a repeated measures structure with several sources of variability for the responses. Moreover, readers and cases are treated as random effects if generalization to a population is required, while tests are clearly fixed effects. We assume that the objective is to compare the accuracy of the tests, however that may be measured (e.g., AUC, PAUC, sensitivity, specificity, etc). Dorfman *et al.* (1992) proposed first transforming the responses to jackknife pseudovalues derived from them, the pseudovalue for a particular case being the difference between the accuracy estimated from all cases and the accuracy estimated from all but that case, and then conducting a mixed-effects ANOVA on the pseudovalues. The null hypothesis is that the mean accuracy of readers is the same for all tests. Song (1997) suggested some variations of this procedure, by treating reader effects as fixed and by matching each case of population P with one from population N so that a pair rather than a single case is omitted when calculating each pseudovalue. However, such matching requires equal numbers of cases in each population, and different matchings may give different outcomes of the analysis. Obuchowski and Rockette (1995) also proposed a mixed-effects ANOVA, but directly on the accuracy values rather than transformed pseudovalues, and including correlated errors in the specification of the model. They identified three distinct correlations in the error terms: for the same reader with different tests, for different readers with the same test, and for different readers with different tests. These correlations were expressed in terms of the different components of variance, hence shown to be estimable from the data, and the usual ANOVA F-tests were modified to take account of them.

On the nonparametric side, Song (1997) extended the approach of DeLong *et al.* (1988) previously described in Section 3.5.1 to handle multiple readers. He developed a Wald statistic for comparing AUC across diagnostic tests, but readers are effectively treated as fixed effects. The remaining nonparametric methods are appropriate for ordinal rather than continuous data so are not strictly within the remit of the present book. However, they merit a mention because they are the only methods that allow the inclusion of covariates in the analysis, and this is valuable if readers' accuracies are thought to be affected by either their own characteristics or those of the patients. The methods are the ordinal regression approach of Toledano and Gatsonis (1996) and the hierachical ordinal regression models of Ishwaran and Gatsonis (2000).

9.6 Further reading

It has been stressed at the start of Section 9.3.1 above that in practical applications the ROC curve is constructed from samples of individuals but conclusions drawn from it are usually applied to the populations from which the samples have been drawn, and that for the conclusions to be valid the samples must be representative of their corresponding populations. This requirement of course applies to statistical investigations in general, and much work has been done over the years on establishing principles of experimental design for ensuring that samples are representative and for maximizing the information that can be extracted from them. An excellent overview of these principles is provided by Cox (1992).

Much work on ROC analysis has been done within radiology, and there are very many relevant articles in radiological journals. Section 9.5 has barely touched on this work, so interested readers are directed to the tutorial reviews/overviews by Obuchowski (2005) and Metz (2008), which between them contain several hundred further references.

Chapter 10

Substantive applications

10.1 Introduction

ROC methodology evolved from practical considerations within the field of radar signal-detection theory (Peterson *et al.*, 1954). Although it drew upon theoretical developments in statistical quality control (Dodge and Romig, 1929; Shewhart, 1931) and statistical inference (Neyman and Pearson, 1933), it has always been firmly rooted in practical applications. Early uses came in psychological testing, but it rapidly became espoused within the field of medical test evaluation where it has found an enormous number of applications over the past twenty years. This wealth of medical examples has already been reflected by many of the illustrative analyses described in earlier chapters, and many more can be found in journals devoted not only to medicine but also to medically-related fields such as radiology, clinical chemistry, cardiology, or health and behavior studies (see, e.g., Collinson, 1998; Kessler, 2002; Obuchowski, 2003; Obuchowski *et al.* 2004). It ought perhaps to be added, though, that while ROC methodology has provided some enormous benefits in these areas, not all ROC studies have been adequately conducted so the interested reader must always be prepared to exercise critical judgement when considering them. In this regard, Obuchowski *et al.* (2004) provide a useful service. They conducted an exhaustive survey of all articles involving ROC curves that appeared in the journal *Clinical Chemistry* in 2001 and 2002, and summarized a number of shortcomings to look out for in similar studies.

That said, the aim of the present chapter is to convey some idea of the breadth of applications to which the theory of the earlier chapters has been put in the literature. In view of the exposure already given to medical applications in preceding illustrative examples, we forgo further mention of them here. Instead, we cover applications in a diverse set of substantive areas as revealed by a moderate (but by no means exhaustive) delve through the literature. Areas covered are atmospheric sciences, geosciences, biosciences, finance, experimental psychology, and sociology. In each case we focus on one or two specific examples in order to give a flavor of some of the uses to which the methodology has been put, and where appropriate provide further references for the interested reader to follow up.

However, we start with one other area, machine learning, which occupies a slightly special niche. It is not a substantive area in the sense of the others listed above, which all relate to measurements made on the world around us and in which data can be collected directly from observations on individuals. It is, instead, concerned with more abstract technicalities associated with computation. On the other hand it comes within computer science, and contains researchers who conduct computational experiments and wish to analyze the results. Moreover, the issues that are addressed relate very centrally to ROC, and have direct bearing on the uses of the methodology in all the other areas. Hence it is appropriate to start our survey with this topic.

10.2 Machine learning

Machine learning is a branch of computer science that is concerned exclusively with induction problems, in which computational algorithms are applied to a given set of data in order to build a model for some descriptive or predictive purpose. We focus specifically on those problems where the model is a classifier that can be used to classify new (i.e., previously unseen) data observations. Standard algorithms for building such classifiers include multi-layer perceptrons, radial basis functions, decision trees, support vector machines, naive Bayes, linear discriminant functions, and logistic regression. The true distribution of data examples and the data generating mechanism are generally assumed to be unknown, so any assessment of performance of the classifier must be based purely on the available data.

Of course, the history of both development and assessment of performance of such classifiers stretches back many years, but when comparing classifiers assessment was predominantly done by estimating

10.2. MACHINE LEARNING

their accuracy of prediction (i.e., by estimating error rate), typically on a number of widely available "benchmark" sets of data. However, Provost *et al.* (1998) argued strongly against this practice when comparing induction algorithms, on several grounds: principally that classification accuracy assumes equal costs of false positive and false negative errors, whereas it is well attested that in real-world problems one type of error can be much more costly than the other; and classification accuracy assumes that the class proportions in the data represent the true prior probabilities of the populations from which they were drawn, whereas this may not be so (in general as well as in particular for the benchmark sets).

Providing that the classification is based on a thresholded value of the classifier, however, a ROC curve can always be constructed and this curve describes the predictive behavior of the classifier independently of either costs or priors (because all possible values of these quantities are represented by varying the classifier threshold). So Provost *et al.* (1998) argued for the use of the ROC curve as the correct means of comparing classifiers. Moreover, they noted that if the ROC curve of one classifier dominated the others (i.e., all other ROC curves were beneath it or at worst equal to it) then that classifier was "best" irrespective of costs or priors, so only in that situation would assessment via classification accuracy be correct. They then conducted some experiments, running a number of induction algorithms on ten datasets chosen from the benchmark UCI repository (Merz and Murphy, 1996) and plotting smoothed ROC curves by averaging the folds of a 10-fold cross-validation. In only one of the datasets was there a dominating smoothed classifier, and they noted that in very few of the 100 runs they performed was there a dominating classifier. They therefore concluded that choosing a classifier from a set of potential classifiers on the basis of accuracy estimates for a given data set was a questionable procedure, the correct comparison between classifiers being on the basis of their ROC curves. Some theoretical results following up these arguments have been established by Nock (2003).

In a similar type of investigation, Bradley (1997) compared four machine learning and two statistical classifiers on six real-world data sets (four taken from the UCI repository, and two others). He found that AUC exhibited a number of desirable properties when compared to accuracy estimation, and showed that the standard deviation of AUC when estimated using 10-fold cross-validation was a reliable estimator of the Mann-Whitney U-statistic. These investigations have a bearing on all empirical studies involving classifier comparison, whatever the substantive area of application.

10.3 Atmospheric sciences

Probabilistic forecasting systems have been much studied since the early 1990s, with particular focus on their validation and accuracy. Such systems attach a probability of occurrence to any event of interest (e.g., precipitation greater than 5 mm, or above-median precipitation). Probability forecasts can be generated in a number of ways, the most common being so-called ensemble predictions in which the probability is obtained as the proportion of a number of constituent members that predict the single dichotomous event. Ensembles can be as simple as a number of independent human forecasters, or as complicated as a number of sophisticated software systems that incorporate diverse technical machinery such as simulation or resampling methodology. For example, two of the latter type of ensemble systems are those of the European Center for Medium-Range Weather Forecasts (ECMRWF) and the National Centers for Environmental Prediction (NCEP; see Buizza et al., 1999).

Whatever the final form of system, if it is employed on n separate occasions there will be a set of n probability values for the event of interest. Moreover, once the time for which the forecast is made has been reached it will be evident whether the event in question has occurred (1) or not (0). So there will be a training set of n pairs of variables, one binary denoting whether the event has occurred or not and the other continuous between 0 and 1 giving the predicted probability s of the event. Now if we are given such an s value for an event, and we want to make a definite prediction as to whether the event will occur or not, then the natural way of doing so is to predict occurrence if s is sufficiently large—that is, if it exceeds some suitable threshold t. So we now have an exactly analogous situation to the ones studied in earlier chapters. Using our previous notation, the cases in which the event has occurred (binary variable value 1) correspond to population P (or "disease" in the medical context) while those where the event has not occurred (binary variable value 0) correspond to population N ("normal"), and the probability variable S is the classifier which allocates an event to population P if $S > t$ and otherwise to population N.

10.3. ATMOSPHERIC SCIENCES

If a definite value is chosen for t (e.g., 0.8) then the accuracy of the 0/1 forecasts can be assessed by comparing observations and predictions using one of the traditional measures (see, e.g., Hand, 1997). More commonly, however, the system is assessed by considering predictions over a range of values of t and producing the ROC curve from the resultant fp and tp values. (Note, however, that in the atmospheric sciences the "R" of ROC is usually taken to be "relative" rather than "receiver.") Having obtained the ROC curve, AUC is generally used as a measure of worth of the system and different systems are compared by means of their AUC values. Moreover, in meteorology there is plenty of scope for further analysis of AUC values. For example, if forecasts are being made on a daily or weekly basis over a period of time then daily or weekly ROC curves can be constructed and a whole time series of AUC values can be obtained. Thus the behavior of different ensemble systems can be compared by comparing these time series.

Some good case studies of this type have been provided by Buizza *et al.* (1999). They conducted statistical analyses of the performance of the ECMWF ensemble system in predicting probability of precipitation. Four seasons were considered between winter 1995 and summer 1997, precipitation was accumulated over 12-hour periods and four events were studied (cumulative precipitation exceeding 1, 2, 5, and 10 mm respectively), and 3-, 5-, 7-, and 10-day forecasts were analyzed in each of three regions (Europe, Alps, and Ireland). This led to many analyses, with ROC curves playing an important role. Support was found for the common belief in atmospheric sciences that for a probabilistic forecast to be useful it should have an AUC value exceeding 0.7. This baseline then enabled time series plots of AUC values to be interpreted, and sample plots illustrated the skill of prediction of the small precipitation amounts (1mm and 2mm) for up to 7 days ahead. One of the main objectives of the study was to evaluate the upgrade to the system that had been introduced in 1996, and the analysis showed that the impact of this upgrade had been substantial with a large gain in predictability. Overall, these case studies provide some nice examples of basic ROC analysis in meteorology.

The atmospheric sciences have also given rise to several other types of curves which are related to ROC curves to a greater or lesser extent. The curve that has the closest relationship is the "relative operating levels" or ROL curve, which is used when the events of interest are associated with some continuous measure of "intensity" and their forecasts are binary (alarm/no alarm), either intrinsically or after conversion from a probability via a fixed threshold. The idea of the ROL curve is again to examine the correct/false alarm relationship, but instead of taking the events as fixed and varying the probability threshold that defines an alarm, the alarm probabilities are taken as fixed and the intensity threshold that defines an event is varied.

To illustrate the idea, consider some data from Mason and Graham (2002) that came from an investigation of different probability forecasts of above-median March-May precipitation over North-East Brazil for 1981-1995. The actual precipitation is expressed as an index of "intensity" in column 2 of Table 10.1, with the median as zero so that the event of interest is associated with positive values of this index. Accordingly, column 3 of the table indicates the events as 1 and the nonevents as 0. Column 4 gives the forecast probabilities of above-median precipitation in each year, derived from a 5-member ensemble system. The worth of this system would be traditionally assessed by a ROC curve in which alarm/no alarm was compared with event/no event for a range of probability thresholds defining an alarm. For the purpose of the ROL curve, however, we take this probability threshold to be fixed, generating an alarm if the probability is greater than or equal to 0.8. Column 5 of the table therefore shows the alarms as 1 and the nonalarms as 0.

The reasoning behind the ROL curve is that if the forecasting system is skillful then greater precipitation should be observed in those cases where an alarm has been issued than in those where there is no alarm. Thus the distribution of event intensity will depend on whether or not an alarm has been issued. Conversely, for a system without skill the distribution of event intensity will be unaffected by presence or absence of an alarm. The ROL curve is a plot of the "correct alarm ratio" (i.e., the proportion of times an event occurred following an alarm) against the "miss ratio" (the proportion of times that an event occurred following no alarm) when the threshold of intensity that defines an event is varied. Thus in this case "alarms" are taken as the "events," "intensity" is taken as the classifier, and an event is predicted when this classifier exceeds the chosen threshold t. The "correct alarm ratio" is therefore just the true positive rate tp and the "miss ratio" is the false positive rate fp of standard ROC theory. To compute the coordinates of points in the empirical ROL curve, we therefore follow the procedure described in Section 3.3, specifically as outlined in Ta-

10.3. ATMOSPHERIC SCIENCES

Table 10.1: Forecasting above-median precipitation in North-East Brazil; data published with permission from the Royal Meteorological Society.

Year	Index	Event/ nonevent	Forecast probability	Alarm/ nonalarm
1981	−1.82	0	0.8	1
1982	−2.33	0	0.8	1
1983	−4.41	0	0.0	0
1984	1.96	1	1.0	1
1985	2.91	1	1.0	1
1986	3.22	1	0.6	0
1987	−0.97	0	0.4	0
1988	2.49	1	0.8	1
1989	3.58	1	0.0	0
1990	−2.28	0	0.0	0
1991	−0.48	0	0.2	0
1992	−3.07	0	0.0	0
1993	−3.46	0	0.0	0
1994	0.12	1	1.0	1
1995	1.5	1	1.0	1

ble 3.1. To do this, we must first order the observations in Table 10.1 in decreasing order of their intensity index values. The first three columns of Table 10.2 give the year, the index value, and the presence or absence of an alarm for this re-ordered set. Then the estimated fp and tp values in columns 4 and 5 are obtained exactly as described in Section 3.3. Plotting these values gives the ROL curve, which is shown along with the chance diagonal in Figure 10.1.

Note that this ROL curve as calculated by Mason and Graham (2002) and given in their Figure 4 has one incorrectly placed point, namely the fourth from the origin; this point should be directly above the third, as in Figure 10.1, and not to its right. Nevertheless, Mason and Graham (2002) calculate the AUC as 0.778, and point out that the same equivalence with the Mann-Whitney U-statistic (Section 3.5.1) holds for AUC in the ROL curve as in the ROC curve. They thus derive a p-value of 0.044 for the significance of departure of the AUC from 0.5,

Figure 10.1: Empirical ROL curve, plus chance diagonal, obtained for the ordered precipitation data in Table 10.2

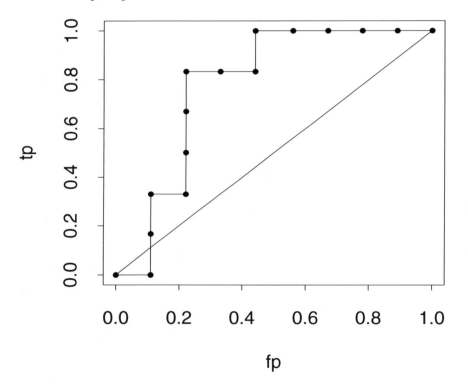

Table 10.2: Re-ordered observations from Table 10.1

Year	Index	Alarm/nonalarm	$\widehat{fp}$	$\widehat{tp}$
1989	3.58	0	0.11	0.0
1986	3.22	1	0.11	0.167
1985	2.91	1	0.11	0.33
1988	2.49	0	0.22	0.33
1984	1.96	1	0.22	0.5
1995	1.5	1	0.22	0.67
1994	0.12	1	0.22	0.833
1991	−0.48	0	0.33	0.833
1987	−0.97	0	0.44	0.833
1981	−1.82	1	0.44	1.0
1990	−2.28	0	0.56	1.0
1982	−2.33	0	0.67	1.0
1992	−3.07	0	0.78	1.0
1993	−3.46	0	0.89	1.0
1983	−4.41	0	1.0	1.0

and conclude that the skill of the forecasting system as measured by this AUC is significantly high. However, despite this significance the AUC value itself is not especially high, reflecting the fact that two nonalarms are attached to the years that have the highest and fourth-highest precipitations respectively.

Another curve that has some connection, albeit more tenuous, with the ROC curve is the "value score" curve developed by Wilks (2001). This author considers the evaluation of probability forecasts for dichotomous events in terms of their economic value over all possible cost/loss ratio decision problems, and develops a value score as a function of the cost/loss ratio. The value score curve is the plot of this score against the cost/loss ratio, and is similar in terms of computational mechanics and graphical display to the ROC curve. It is used to compare objective statistical forecasts from Model Output Statistics with subjective probability forecasts made by US National Weather Service forecasters, and it is shown that there are ranges of cost/loss ratios for which each forecast source is the more valuable. Additionally, it is shown that the value score curve is sensitive to either conditional or unconditional biases whereas the ROC curve is not; so Wilks (2001) argues that the latter curve should be viewed as reflecting potential rather than actual skill.

Readers who are interested in further applications of ROC methodology in atmospheric sciences will find about 20 additional references to relevant studies in the article by Mason and Graham (2002).

10.4 Geosciences

The derivation of models for the prediction of natural hazards is a major concern in the earth sciences. Various such hazards have been studied, probably the two most common being landslides (Brenning, 2005) and earthquakes (Holliday et al., 2006). Prediction of landslides is principally a spatial problem, the aim being to use terrain parameters, geological attributes, and environmental conditions to identify areas that are susceptible to future landsliding. By contrast, prediction of earthquakes is principally a temporal problem, the area under threat often being already well defined and the main objective being to predict the likely time interval between earthquakes. However, the common feature in these two types of hazard, shared with the methods already discussed in the atmospheric sciences, is the fact that the predictions are often probabilistic and this leads to the use of ROC methodology for assessing the worth of the predictive models.

In the case of landslide data, the training set typically consists of map grid points of known class membership (i.e., landslide or non-landslide) and associated covariates such as elevation, slope, aspect, curvature, distance to ridge, vegetation, and infrastructure, among others. In principle there are many potential classifiers that could form the basis of the prediction, but a problem with spatial data is that the grid-point observations are not statistically independent because of their spatial proximity. This spatial (auto)correlation violates the independence requirement of many statistical modeling procedures, and hence limits the potential pool of classifiers to those that can be adapted to cope with spatial autocorrelation. Generalized linear models, in particular logistic regression, can be so adapted (Augustin et al., 1996; Gotway and Stroup, 1997). Moreover, routines for performing the model fitting are readily available in Splus or R (Venables and Ripley, 2002), so logistic regression is a popular classifier of choice for landslide data (see Table 1 of Brenning, 2005). Once a logistic model has been fitted to the training data, the probability of a landslide can be obtained directly for any given set of covariate values and a binary prediction (landslide/no landslide) can be obtained by applying a threshold cut-off to the calculated probability.

Turning to assessment of the fitted classifier, various problems can be identified. Assessment must be conducted on separate data from the training set that has been used in the fitting of the classifier, and various strategies are possible when seeking such a test set. It could come from the training area but at a different time, or from an adjacent area. However, landslide causes and characteristics may vary systematically in space or time, in which case the computed predictions will relate to different distributions from the ones that govern the training data. A common way of circumventing such problems is to choose a random subset of the training data, set it aside, and use it as the test data. Many grid-point sets of data are sufficiently large (upwards of 500,000 cells) for this to be perfectly feasible, but spatial dependencies between training and test data points that are close together may bias the predictions optimistically. To guard against all these possibilities, it is advisable to base estimates on a number of replicates of randomly chosen test sets. Moreover, one can isolate "intra-domain" error rates (test samples chosen from inside the training area) from "extra-domain" error rates (test samples chosen from the edges of the training area) for reporting purposes. Estimates of tp and fp obtained in this way over a range of threshold values thus enable the empirical ROC curve to be constructed, and the estimated value of AUC provides a measure of worth of the predictive model. For further details of some case studies of this type, see Gorsevski *et al.* (2000) or Ayalew and Yamagishi (2005).

By contrast with landslides, Holliday *et al* (2006) study the prediction of major earthquakes in California. Their database consists of a coarse-grained mesh of the region into pixels of side approximately 11 km in length. The average intensity of earthquake activity $I(\boldsymbol{x}, t)$ at a point $\boldsymbol{x}$ up to year (time) t is defined to be the number of earthquakes (Richter scale ≥ 6) that have occurred in the region enclosed by the pixel centered at $\boldsymbol{x}$ between 1932 (when records began) and t. Let $S(t)$ denote the total number of earthquakes in the whole region up to time t, that is, the sum of $I(\boldsymbol{x}, t)$ over all $\boldsymbol{x}$. Then a simple probability $P_1(\boldsymbol{x})$ for the occurrence of an event at $\boldsymbol{x}$ in the future (i.e., at time $t_1 > t$)

is given by $I(\boldsymbol{x}, t)/S(t)$. A second probability measure $P_2(\boldsymbol{x})$ can be derived as the average squared change in intensity over a specified time interval, and Holliday et al. (2006) follow previous authors in using an interval of 13 years. This probability measure is somewhat more complicated than P_1, due to the more intricate computations needed to obtain the intensity change, so the details are omitted here and the reader is referred to Holliday et al. (2006). These authors call P_1 the "average map" and P_2 the "fluctuation map."

A binary prediction (earthquake/no earthquake) can be obtained as usual by thresholding either of these probability predictions at an appropriate level d. The predictive power of the chosen method can be assessed by dividing the time interval over which the data have been gathered into two portions, (t_0, t_1) and (t_1, t_2) where t_0 is the start year, forming predictions from the data in the first portion, and estimating tp and fp rates by comparing the predictions with the actual occurrences in the second portion. ROC curves for each probability measure P_1 and P_2 are constructed by varying the thresholds d and plotting the resulting tp and fp values. The two measures can then be compared by means of the AUC_1 and AUC_2 under their respective curves. Note, however, that these areas are known in the geoscience literature as the "Pierce skill scores." Moreover, a summary measure often computed in these studies is the Ginzburg criterion given by the ratio of the areas, $\mathcal{G} = \text{AUC}_2/\text{AUC}_1$, the fluctuation map providing a better (poorer) forecast than the average map if $\mathcal{G} > 1$ (< 1). Holliday et al. (2006) analyzed the data using varying time intervals for the computation of P_2, and found that since 1960 major earthquakes in California tend to occur preferentially during intervals of time when $\mathcal{G} < 1$. They noted that this finding was consistent with mean field dynamics.

Further ROC applications in geosciences arise in a number of areas. For example, Goovaerts et al. (2005) consider methods for the detection of anomalies in data from high spatial resolution hyperspectral (HSRH) imagery and give references to a number of other studies with such data, while Thuiller et al. (2003) employ ROC methodology in the study of forest tree distributions.

10.5 Biosciences

Microarray technology has taken a firm root in the biosciences in a very short space of time, and much interest is consequently directed at the analysis of microarray data. A single microarray experiment allows the measurement of expression levels in cells of organisms for thousands or even tens of thousands of genes. Each such measurement is usually referred to as a "biomarker." Frequently, such experiments are geared towards measuring differences between two groups of individuals, for example two species of insects, two varieties of plants, organisms measured under two different atmospheric conditions, or diseased versus normal individuals. Generally, there is an implied objective of providing a means of predicting the true group of a "new" individual, so the framework is that of classical discriminant analysis. The problem is that the large cost and effort of conducting such experiments means that a typical microarray data set has only tens of observations (individuals) for each of which there might be tens of thousands of biomarkers (variables). Such heavily unbalanced data sets produce many problems, both computational and statistical, in their analysis, and consequently there is much current research devoted to overcoming these problems. We give an outline here of two strands of this research, both of which involve ROC methodology.

Ma and Huang (2005) and Ma et al. (2006) tackle the problem of deriving an efficient classifier to allocate future individuals to one or other group. As usual, let us denote the populations from which the individuals have come by N and P, the vector of measurements (biomarkers) for each individual by $\boldsymbol{X} = (X_1, X_2, \ldots, X_p)^T$, and the classifier for allocating individuals to N or P by $S(\boldsymbol{X})$. Ma and Huang (2005) propose a monotone increasing function of a linear combination of the biomarkers as the classifier, that is

$$S(\boldsymbol{X}) = G(\beta_1 X_1 + \beta_2 X_2 + \ldots + \beta_p X_p)$$

for some (unknown) monotonic function $G(\cdot)$ and parameters $\beta_1, \beta_2, \ldots \beta_p$. The benefit of restricting this function to be monotonic is that the classification rule can be reduced to one that classifies to P or N according as the value of $\beta_1 X_1 + \beta_2 X_2 + \ldots + \beta_p X_p$ is greater than or less than some constant c.

Thus, for any given value of $\boldsymbol{\beta} = (\beta_1, \beta_2, \ldots \beta_p)^T$ applied to the microarray data, we can construct the empirical ROC curve and calculate its area AUC using the methods of Sections 3.3 and 3.5. But what is the best value to use for $\boldsymbol{\beta}$? In view of the lack of any distributional assumptions from which to build a probability model, and the lack of dependence on the function G, Pepe et al. (2006) argue for the value that gives the "best" ROC curve, namely the one that maximizes its area AUC. This turns out to be a special case of the maximum rank correlation (MRC) estimator of Han (1987), and is the one adopted by Ma and Huang (2005). However, since the empirical AUC is not continuous while the microarray is high-dimensional (i.e., p is very large), finding the maximum presents severe computational problems. These can be avoided by fitting a smooth function to the empirical AUC, so Ma and Huang approximate it by the sigmoid function $1/[1 + \exp(-x/\epsilon)]$ where ϵ is a small data-dependent positive number included to avoid bias near $x = 0$. Also, for identifiability, $|\beta_1|$ is set to 1. A rule of thumb is given for determining ϵ, and an algorithm is presented for carrying out the MRC estimation process. The latter involves a tuning parameter, so a v-fold cross-validation scheme is also outlined for setting this parameter.

Ma and Huang (2005) present some numerical simulation studies to demonstrate that the estimation process is satisfactory and that the patterns in the AUCs for varying sample sizes are as would be expected. They also analyze several sets of data, one with 2000 biomarkers on 62 individuals and the other with 7129 biomarkers on 49 individuals. In each case, the number of biomarkers was first reduced to 500 to ensure computational stability, the chosen ones being the 500 that had the largest absolute values of the adjusted t-statistics between the two groups. The adjustment to the t-statistics is a shrinkage-type adjustment which guards against small-sample volatility, and by choosing the biomarkers that have the largest absolute t-statistic values one can be confident that most or all the "important" biomarkers are present. Ma et al. (2006) refine the method by fitting a binormal ROC model rather than the empirical ROC with sigmoid smoothing of the AUC, and show that this is computationally more affordable as well as being robust in small sample cases.

The question as to which of the biomarkers are "important" in distinguishing between the groups is a central one, particularly as there are usually very many biomarkers but relatively few will be expected to be important while the rest are "noisy." The choice of 500 as has been done above is perhaps a partial solution but by no means an optimal one: high correlations between biomarkers will lead to redundancies in the set, there will still be some (or many) noise variables retained, and there is no guarantee that important ones haven't been lost. In fact, the selection on the basis of t-statistic values is an example of a "one-gene-at-a-time" approach (Dudoit and Fridlyand, 2003), and could equally well have been conducted on the basis of nonparametric statistics, p-values or information gain. Mamitsuka (2006) briefly reviews this and other existing filtering approaches (Markov blanket filtering, Xing filtering) before proposing a new method that makes use of some of these approaches but is based on the ROC curve.

In Mamitsuka's approach, the first step is to compute and rank the AUC values when each biomarker is used on its own. This is therefore a "one-gene-at-a-time" step, but using AUC as the criterion instead of those above and hence more likely to produce relevant selections. Moreover, Mamitsuka provides formulae for rapid calculation of these AUC values from ordered biomarker values. An initial set of important biomarkers comprises the ones having the top AUC values. From this initial selection, redundant biomarkers are iteratively eliminated using a Markov blanket based on ROC curves until the optimal set is achieved. Mamitsuka (2006) gives the full algorithmic specification, and describes a number of simulation experiments that show this method outperforming competitors on a range of data sets.

Further uses of ROC in the biosciences can be found by following up references in the articles cited above.

10.6 Finance

Financial institutions such as banks, when considering whether to offer personal loans to individuals, want to protect themselves against loss due to customers defaulting on their repayments. Consequently, a *credit scoring* technology has been developed, whereby the institution will gather information on a number of variables relating to income, expenditure, savings, and behavior for each individual applying for a loan and combine this information into a single score on the basis of which the loan is either granted or refused. We are thus in exactly the same position as for many of the preceding examples. From the bank's perspective there are two populations of individuals, "defaulters" (P) and "nondefaulters" (N); there is a vector $\boldsymbol{X}$ of values that can be gathered for each applicant; and there is a credit scoring model $S(\boldsymbol{X})$ that converts this vector of values into a single score such that the application is rejected if the score exceeds a given threshold t and granted otherwise. As before, the score can be the (estimated) probability that the applicant will default or it can be just some arbitrary scale of values, but by the usual convention a high score implies low creditworthiness. Various credit scoring models are now available to the financial institutions, and they can be fitted and calibrated on data that the institution has built up from its database of past records of X values for known defaulters and nondefaulters. If an institution wishes to find the best model to use then ROC and AUC can be employed in the usual way.

The main practical task for the institution is to determine an appropriate threshold value t for the chosen credit scoring model. Blöchlinger and Leippold (2006) note that this is an arbitrary choice, most often based on qualitative arguments such as business constraints, and so will generally be suboptimal. They argue that a more rigorous criterion can be derived from knowledge of the prior probability of default along with the associated costs and revenues, so use the sort of development shown in Section 2.3.1 to arrive at an optimal choice. This choice is given by the point at which the line with slope s forms a tangent to the ROC curve, where s is given by equation (6) of Blöchlinger and Leippold (2006) as a function of the prior probabilities, the revenue, the loss given default, and the strategic value. They illustrate the optimal choice over different cash-flow scenarios, by showing the ROC curves and the lines of slope s for varying costs and revenues. They also consider pricing regimes where the institution sets the price of the loan according to the credit score, and demonstrate the link between pricing rule and ROC curve.

A different aspect of personal loan data is considered by Thomas *et al.* (1999) and by Stepanova and Thomas (2002). Instead of being concerned simply with whether the customer will default or not, they focus on profitability to the institution in which case time becomes important. If a customer defaults but the time to default is long, then the acquired interest may have exceeded the cost of default and the institution has profited. Conversely if the customer pays off the loan early, then the lender may lose a proportion of the interest. Because time is now involved, and data sets gathered by institutions are necessarily snapshots at limited periods of time, information is not necessarily available on all defaulters and how long it took for them to default. Thus some observations are *censored*, and methods of *survival analysis* need to be used when attempting to draw inferences from the data. If interest focusses on time to default, then all loans defaulted in the database are considered to be "failures" and all the others are considered as censored (on the grounds that they might default in the

10.6. FINANCE

future). Likewise, if interest focusses on early repayment then the early repayers replace the defaulters in this framework. Fitting one of the standard survival models to the data then provides a mechanism for predictions.

Various survival models have been proposed and studied, such as the Weibull, the exponential, and Cox's proportional hazards model; the last of these is by far the most popular in practical applications. The different models specify different forms for either the *survivor function* $S(t)$, which is the probability that failure will occur after time t, or for the *hazard function* $h(t)$, which is the probability of a failure at time t given that the individual has survived up to that time. Standard theory shows that the two functions are connected through the relationship

$$S(t) = \exp\left(-\int_0^t h(u)du\right).$$

If we have covariates $x_1, x_2, \ldots, x_p$ whose effect on time to failure we want to assess, Cox (1972) proposed the model

$$h(t; x_1, x_2, \ldots, x_p) = e^{\beta_1 x_1 + \beta_2 x_2 + \ldots + \beta_p x_p} h_0(t),$$

where the β_i are parameters and h_0 is an unknown function giving the baseline hazard when all the x_i are zero. The attractive feature of this model is that the parameters can be estimated without any knowledge of $h_0(t)$, just by using rank of failure and censored times. Stepanova and Thomas (2002) summarize the essential technical details for application to personal loan data. They then apply the method to a data set comprising data on 50,000 personal loan applications, with 16 characteristics recorded for each application along with the repayment status for each month of the observation period of 36 months.

As well as fitting Cox's model they compare it with a logistic regression approach, separately for the cases of early repayment and defaulting. Comparisons are mainly effected using ROC curves, although there are also some more detailed and specific comparisons. As a broad statement of conclusions, there does not seem to be much to choose between the methods when the covariates are treated as continuous, the logistic regression approach having the slight edge. But when the covariates are treated as categorical by segmenting them (the more usual scenario in financial applications where the natural unit is the month), then Cox's model shows distinct advantages particularly in the case of early repayment. Extending the model by including time-by-covariate interactions gives some further improvements, as it allows the effects of application characteristics to either increase or decrease with age of the loan.

10.7 Experimental psychology

A number of different areas of Experimental Psychology provide scope for ROC methodology when analyzing the results. One such area is that of item recognition, a common line of study being one in which the ability of individuals to recognize previously memorized items is subjected to various memory enhancing or inhibitory mechanisms and their recall success rates are compared using ROC curves. We here briefly consider two recent instances of such studies.

Verde and Rotello (2004) considered the "revelation effect," which is the increased tendency to "recognize" items when the judgement is preceded by an incidental task, using a number of linked experiments. The basic experimental setup was as follows. Each participating subject was first presented sequentially with 50 words to learn, of which the middle 40 were critical for the experiment and the remaining 10 were "fillers." These words were chosen randomly from a pool of 300 low-frequency eight-letter words, and each subject was given several seconds in which to remember them. Each subject was then presented with a further sequential list of 80 words, half of which were from the "learned" list (i.e., "old") and the other half were new. The subject had to mark a 10 centimeter line, one end of which was labeled *very sure old* and the other *very sure new*, indicating how confident the subject was that the presented word was one that had been learned. However, prior to half of the 80 words, an anagram was presented which the subject had to unscramble using an anagram key; the anagram appeared equally often before new words as before old words. In the first experiment (*unrelated revelation*) the anagrams were of words unrelated to the "old" list; in the second experiment (*identical revelation*) the anagram was of the word that was then presented for recognition; in the third experiment *all* 80 words were preceded by an anagram, half being unrelated and the other half the recognition words, while a fourth experiment had all three possibilities (no anagram, unrelated anagram, and identical anagram).

The classification score for each subject and each recognition task is just the distance along the 10 cm line in the direction of "very sure old," so that a binary choice between old and new can be deduced by thresholding this distance. Once again, therefore, the setup is conducive to ROC analysis, and Verde and Rotello (2004) produced ROC curves for each of the conditions represented in each of the experiments. The conditions were unrelated anagrams/no anagrams, identical anagrams/no anagrams, unrelated/identical anagrams, and no/unrelated/identical anagrams for the four experiments respectively. Inspection of these curves showed no material difference between unrelated anagrams and no anagrams, a substantial difference between identical anagrams and no anagrams, and a small difference between unrelated and identical anagrams, with identical anagrams degrading the quality of the classifiers. In the discussion, Verde and Rotello (2004) speculated as to the reasons behind this finding.

Algarabel and Pitarque (2007) likewise conducted experiments into word recognition, but in addition included context recognition via combinations of pairs of colors against which the words were displayed. They fitted ROC curves, and then analyzed three parameters of the fitted curves, namely the curvature of the untransformed curves and the linearity and curvature of the normally-transformed curves. This paper contains a number of references to ROC analysis in related studies, so can be used as a springboard for further reading.

10.8 Sociology

The intersection of sociology, criminology, and behavioral psychology is an area in which many different scoring systems have been proposed in order to derive quantitative measures for predicting transgressive behavior of some sort. These scoring systems go under various different titles, such as behavior scales, personality assessment inventories, risk assessment measures or screening indices, and the behavior that they are intended to predict might range from highly criminal action to lighter matters such as propensity for absenteeism from work. Much effort has been expended on assessment and comparison of these scoring systems in order to determine which are good ones for general use, and inevitably ROC analysis has formed a central technique in these studies. However, the ROC methodology used has generally been fairly straightforward, so it does not warrant extensive description and we here just briefly outline a few of the studies in order to give a flavor of their breadth.

Hayes (2002) took as her starting point the fact that people with an intellectual disability are significantly over-represented in many Western criminal justice systems, making early identification of such disability important if the individuals are to receive adequate attention. In order to identify such individuals she developed the Hayes Ability Screening Test, and compared it against the existing Kaufman Brief Intelligence Test and the Vineland Adaptive Behavior Scales by testing a sample of 567 adult and juvenile offenders, in custodial and community settings. ROC analysis was used to examine the effectiveness with which each scale discriminated between people with and without intellectual disability. The Hayes test showed a high AUC of 0.870, with an optimum threshold value of 85 for the test score that yielded 82.4% accuracy in detecting true positives and 71.6% accuracy in excluding true negatives when compared with the Kaufman test. In comparison with the Vineland scale these percentages were 62.7 and 71.2.

Craig et al. (2006) were concerned with predicting reconviction among sexual and violent offender groups, and specifically wished to evaluate the predictive accuracy of two new risk measures as well as comparing them with four existing ones. They conducted the investigation on a sample of 85 sexual, 46 violent, and 22 general offenders in a 10-year follow-up to establish reconviction patterns. The assessment and comparison of the measures involved cross-validation and ROC curve analysis, specifically via AUC. Inspection of the table of AUC values for all combinations of risk measure and offense reconviction at 2-year, 5-year, and 10-year follow-up showed that the new measures performed very well against the existing ones with AUC peaking at 0.87 for some reconviction categories.

Finally, Kucharski et al. (2007) assessed the Personality Assessment Inventory (PAI) Validity scales for the detection of malingered psychiatric disorders. Having started with 380 criminal defendants in Atlanta, Georgia as potential participants in the study, and having excluded individuals following results of several psychological tests, they ended up with 116 participants. These were divided into malingering and nonmalingering groups on the basis of their performance on the Structured Interview of Reported Symptoms (SIRS) scale, a reliable measure of malingering that has been validated in many studies. ROC analysis was then used to investigate several of the PAI scales, and demonstrated acceptable sensitivity and specificity for one of these scales (the Negative Impression Management scale) but not the others.

Appendix: ROC Software

Many of the researchers who have developed the methodology described in this book have also written software to perform the relevant analyses, and have made this software available via free downloads from the world wide web. Here are just a few examples of such websites. In each case we give the home institution of the site, the lead researcher associated with the software, and some brief notes about the content of the site.

1. Cleveland Clinic, Department of Quantitative Health Sciences; Obuchowski and colleagues.
   http://www.bio.ri.ccf.org/html/rocanalysis.html
   This site gives a nice overview of the various methodological aspects, advice on which programs to use for different objectives and data features, a large set of downloadable programs, and many links to other ROC sites (including those listed below). Two closely related sites for specific applications are:
   http://www.bio.ri.ccf.org/html/obumrm.html for a FORTRAN program to implement the Obuchowski & Rockette method for MRMC data; http://www.bio.ri.ccf.org/html/ractivities.html for a link to nonparametric ROC analysis.

2. Fred Hutchinson Cancer Center (University of Washington), Diagnostics and Biomarkers Statistical (DABS) Center; Pepe and colleagues.
   http://labs.fhcrc.org/pepe/dabs/
   This site also has many downloadable programs, classified according to the language or statistical package (Stata, R/Splus, SPSS, SAS, and FORTRAN) for which they are appropriate. There are some additional links, including one to the Cleveland site above and another to the commercial package MedCalc.

3. University of Chicago, Kurt Rossman Laboratories; Metz and colleagues.
   http://www-radiology.uchicago.edu/krl/rocstudy.htm
   This site specializes in ROC analysis in Medical Imaging. The long-standing focus has been on maximum-likelihood estimation of ROC curves and on testing the statistical significance of differences between ROC estimates. Brief summaries of the methodology are given for each of the programs.

4. University of Iowa, Department of Radiology, Medical Image Perception Laboratory; Berbaum and colleagues.
http://perception.radiology.uiowa.edu/
Some of the programs have been developed in collaboration with the Chicago group above.

The above sites are primarily concerned with medical applications. Workers in data mining and machine learning can gain access to relevant software on

http://www.kdnuggets.com/software/classification-analysis.html

while readers particularly interested in exploring ROC facilities in either R or MATLAB® are directed respectively to

http://rocr.bioinf.mpi-sb.mpg.de/ for R, and
http://theoval.sys.uea.ac.uk/matlab/default.html#roc for MATLAB®.

If a basic analysis of a small data set is required, then this can be carried out very conveniently online using the ROC calculator provided by John Eng on the John Hopkins site

http://www.rad.jhmi.edu/jeng/javarad/roc/JROCFITi.html.

Finally, users looking for an approved commercial ROC package may be interested in the review of 8 programs that was conducted by Stephan et al. (2003). These authors rated each package on a number of features (data input, data output, analysis results, program comfort and user manual), but from the results were only able to positively recommend three packages:

Analyse-it (www.analyse-it.com),
AccuROC (www.accumetric.com), and
MedCalc (www.medcalc.be).

However, as a cautionary note they pointed out that no single program fulfilled all their expectations perfectly, and every program had disadvantages as well as advantages. The cited reference gives full particulars.

References

Adams, N.M and Hand, D.J. (1999). Comparing classifiers when the misallocation costs are uncertain. *Pattern Recognition*, **32**, 1139-1147.

Algarabel, S. and Pitarque, A. (2007). ROC parameters in item and context recognition. *Psichothema*, **19**, 163-170.

Allison, P.D. (2001). *Missing Data*. Sage Publications, Thousand Oaks, CA.

Alonzo, T.A. and Pepe, M.S. (2002). Distribution-free ROC analysis using binary regression techniques. *Biostatistics*, **3**, 421-432.

Alonzo, T.A. and Pepe, M.S. (2005). Assessing accuracy of a continuous screening test in the presence of verification bias. *Applied Statistics*, *54*, 173-190.

Alonzo, T.A., Pepe, M.S. and Lumley, T. (2003). Estimating disease prevalence in two-phase studies. *Biostatistics*, **4**, 313-326.

Augustin, N.H., Mugglestone, M.A. and Buckland, S.T. (1996). An autologistic model for the spatial distribution of wildlife. *Journal of Applied Ecology*, **33**, 339-347.

Ayalew, L. and Yamagishi, H. (2005). The application of GIS-based logistic regression for landslide susceptibility mapping in the Kakuda-Yahiko mountains, Central Japan. *Geomorphology*, **65**, 15-31.

Bamber, D. (1975). The area above the ordinal dominance graph and the area below the receiver operating characteristic graph. *Journal of Mathematical Psychology*, **12**, 387-415.

Begg, C.B. (1989). Experimental design of medical imaging trials: issues and options. *Investigative Radiology*, **24**, 934-936.

Begg, C.B. and Greenes, R.A. (1983). Assessment of diagnostic tests when disease is subject to selection bias. *Biometrics*, **39**, 207-216.

Beiden, S.V., Campbell, G., Meier, K.L. and Wagner, R.F. (2000).

On the problem of ROC analysis without truth. In: *Proceedings of the Society of Photo-Optical Instrumentation Engineers (SPIE)*, eds. M.D. Salman, P.Morley and R. Ruch-Gallie, vol 3981, pp. 126-134. The International Society for Optical Engineering, Bellingham, WA.

Biggerstaff, B.J. and Tweedie, R.L. (1997). Incorporating variability in estimates of heterogeneity in the random effects model in meta-analysis. *Statistics in Medicine*, **16**, 753-768.

Bloch, D.A. (1997). Comparing two diagnostic tests against the same "gold standard" in the same sample. *Biometrics*, **53**, 73-85.

Blöchlinger, A. and Leippold, M. (2006). Economic benefit of powerful credit scoring. *Journal of Banking and Finance*, **30**, 851-873.

Bradley, A.P. (1997). The use of the area under the ROC curve in the evaluation of machine learning algorithms. *Pattern Recognition*, **30**, 1145-1159.

Branscum, A.J., Gardner, I.A. and Johnson, W.O. (2005). Estimation of diagnostic-test sensitivity and specificity through Bayesian modeling. *Preventive Veterinary Medicine*, **68**, 145-163.

Brenning, A. (2005). Spatial prediction models for landslide hazards: review, comparison and evaluation. *Natural Hazards and Earth System Sciences*, **5**, 853-862.

Briggs W.M. and Zaretzki R. (2008). The skill plot: a graphical technique for evaluating continuous diagnostic tests. *Biometrics*, **63**, 250-261.

Brooks, S.P. (1998). Markov chain Monte Carlo method and its application. *The Statistician*, **47**, 69-100.

Brumback, L.C., Pepe, M.S. and Alonzo, T.A. (2006). Using the ROC curve for gauging treatment effect in clinical trials. *Statistics in Medicine*, **25**, 575-590.

Buizza, R., Hollingsworth, A., Lalaurette, F. and Ghelli, A. (1999).

Probabilistic predictions of precipitation using the ECMWF ensemble prediction system. *Weather and Forecasting*, **14**, 168-189.

Cai, T. and Moskowitz, C.S. (2004). Semi-parametric estimation of the binormal ROC curve for a continuous diagnostic test. *Biostatistics*, **5**, 573-586.

Cai, T. and Pepe, M.S. (2003). Semi-parametric ROC analysis to evaluate biomarkers for disease. *Journal of the American Statistical Association*, **97**, 1099-1107.

Cai, T. and Zheng, Y. (2007). Model checking for ROC regression analysis. *Biometrics*, **63**, 152-163.

Campbell, G. and Ratnaparkhi, M.V. (1993). An application of Lomax distributions in receiver operating characteristic (ROC) curve analysis. *Communications in Statistics*, **22**, 1681-1697.

Chi, Y.-Y. and Zhou, X.-H. (2008). Receiver operating characteristic surfaces in the presence of verification bias. *Applied Statistics*, **57**, 1-23.

Choi, Y.-K., Johnson, W.O., Collins, M.T. and Gardner, I.A. (2006). Bayesian inferences for Receiver Operating Characteristic curves in the absence of a gold standard. *Journal of Agricultural, Biological and Environmental Statistics*, **11**, 210-229.

Cochran, W.G. (1964). Comparison of two methods of handling covariates in discriminant analysis. *Annals of the Institute of Statistical Mathematics*, **16**, 43-53.

Cochran, W.G. (1977). *Sampling Techniques* (3rd Edition). Wiley, New York.

Cochran, W.G. and Bliss, C.I. (1948). Discriminant functions with covariance. *Annals of Mathematical Statistics*, **19**, 151-176.

Coffin, M. and Sukhatme, S. (1995). A parametric approach to measurement errors in receiver operating characteristic studies. *Lifetime Data Models in Reliability and Survival Analysis*, 71-75. Kluwer, Boston.

Coffin, M. and Sukhatme, S. (1997). Receiver operating characteristic studies and measurement errors. *Biometrics*, **53**, 823-837.

Collinson, P. (1998). Of bombers, radiologists, and cardiologists: time to ROC. *Heart*, **80**, 215-217.

Cortes, C. and Mohri, M. (2005). Confidence intervals for the area under the ROC curve. *Advances in Neural Information Processing Systems vol 17 (NIPS 2004)*. MIT Press, Cambridge, MA.

Cox, D.R. (1972). Regression models and life-tables (with discussion). *Journal of the Royal Statistical Society, Series B*, **74**, 187-220.

Cox, D.R. (1992). *Planning of Experiments*. Wiley, New York.

Craig, L.A., Beech, A.R. and Browne, K.D. (2006). Evaluating the predictive accuracy of sex offender risk assessment measures on UK samples: a cross-validation of the risk matrix 2000 scales. *Sexual Offender Treatment*, 1(1). (Electronic Journal, http://www.sexual-offender-treatment.org/19.98)

Davis J. and Goadrich M (2006). The relationship between precision-recall and ROC curves. In: *Proceedings of the 23rd International Conference on Machine Learning (ICML)*, ed. W.W.Cohen and A.Moore, 233-240.

Davison, A.C. and Hinkley, D.V. (1997). *Bootstrap Methods and their Applications*. Cambridge University Press, Cambridge.

DeLong, E.R., DeLong, D.M. and Clarke-Pearson, D.L. (1988). Comparing the areas under two or more correlated receiver operating characteristic curves: a nonparametric approach. *Biometrics*, **44**, 837-845.

Dendurki, N. and Joseph, L. (2001). Bayesian approaches to modelling the conditional dependence between multiple diagnosis tests. *Biometrics*, **57**, 158-167.

Denison, D.G.T, Holmes, C.C., Malik, B.K. and Smith, A.F.M. (2002).

Bayesian Methods for Nonlinear Classification and Regression. Wiley, Chichester.

DerSimonian, R. and Laird, N. (1986). Meta-analysis in clinical trials. *Control Clinical Trials*, **7**, 177-188.

Dodd, L.E. and Pepe, M.S. (2003a). Semiparametric regression for the area under the receiver operating characteristic curve. *Journal of the American Statistical Association*, **98**, 409-417.

Dodd, L.E. and Pepe, M.S. (2003b). Partial AUC estimation and regression. *Biometrics*, **59**, 614-623.

Dodge, H. and Romig, H. (1929). A method of sampling inspection. *Bell Systems Technical Journal*, **8**, 613-631.

Dorfman, D.D. and Alf, E. (1969). Maximum likelihood estimation of parameters of signal detection theory and determination of confidence intervals—rating method data. *Journal of Mathematical Psychology*, **6**, 487-496.

Dorfman, D.D., Berbaum, K.S. and Metz, C.E. (1992). Receiver operating characteristic rating analysis: generalization to the population of readers and patients with the jackknife method. *Investigative Radiology*, **27**, 307-335.

Dorfman, D.D., Berbaum, K.S., Metz, C.E., Lenth, R.V., Hanley, J.A. and Dagga, H.A. (1997). Proper receiver operating characteristic analysis: the bigamma model. *Academic Radiology*, **4**, 138-149.

Dreiseitl, S., Ohno-Machado, L., and Binder, M. (2000). Comparing three-class diagnostic tests by three-way ROC analysis. *Medical Decision Making*, **20**, 323-331.

Drummond, C. and Holte, R.C. (2000). Explicitly representing expected cost: an alternative to ROC representation. In: *Proceedings of the 6th ACM SIGKDD International Conference on Knowledge Discovery and Data Mining*, 198-207.

Dudoit, S. and Fridlyand, J. (2003). Classification in microarray experiments. In: *Statistical Analysis of Gene Expression Microarray Data*, ed. T. Speed, pp. 93-158. Chapman & Hall, London.

Edwards, D.C., Metz, C.E., and Kupinski, M.A. (2004). Ideal observers and optimal ROC hypersurfaces in N-class classification. *IEEE Transations on Medical Imaging*, **23**, 891-895.

Edwards, D.C., Metz, C.E., and Nishikawa, R.M. (2005). The hypervolume under the ROC hypersurface of "near-guessing" and "near-perfect" observers in N-class classification tasks. *IEEE Transactions on Medical Imaging*, **24**, 293-299.

Efron, B. and Tibshirani, R.J. (1993). *An Introduction to the Bootstrap*. Chapman & Hall/CRC, New York.

Egan, J.P. (1975). *Signal Detection Theory and ROC Analysis*. Academic Press, New York.

Enøe, C., Georgiadis, M.P. and Johnson, W.O. (2000). Estimation of sensitivity and specificity of diagnostic tests and disease prevalence when the true disease state is unknown. *Preventive Veterinary Medicine*, **45**, 61-81.

Erkanli, A., Sung, M., Costello, E.J. and Angold, A. (2006). Bayesian semi-parametric ROC analysis. *Statistics in Medicine*, **25**, 3905-3928.

Escobar, M.D. and West, M. (1995). Bayesian density estimation and inference using mixtures. *Journal of the American Statistical Association*, **90**, 577-588.

Everson, R.M. and Fieldsend, J.E. (2006). Multi-class ROC analysis from a multi-objective optimisation perspective. *Pattern Recognition Letters*, **27**, 918-927.

Faraggi, D. (2000). The effect of random measurement error on receiver operating characteristic (ROC) curves. *Statistics in Medicine*, **19**, 61-70.

REFERENCES

Faraggi, D. (2003). Adjusting receiver operating characteristic curves and related indices for covariates. *Journal of the Royal Statistical Society, Series D*, **52**, 179-192.

Faraggi, D. and Reiser, B. (2002). Estimation of the area under the ROC curve. *Statistics in Medicine*, **21**, 3093-3106.

Fawcett, T. (2006). An introduction to ROC analysis. *Pattern Recognition Letters*, **27**, 861-874.

Ferguson, T.S. (1983). Bayesian density estimation by mixtures of normal distributions. In: *Recent Advances in Statistics*, eds. H. Rizvi and J. Rustagi, pp. 287-302. Academic Press, New York.

Ferri C., Hernández-Orallo J., and Salido M.A. (2004). Volume under the ROC surrface for multi-class problems. *Lecture Notes in Computer Science*, **2837**, 108-120.

Fisher, R.A. (1936). The use of multiple measurements in taxonomic problems. *Annals of Eugenics*, **7**, 179-184.

Fleiss, J.L. (1981). *Statistical Methods for Rates and Proportions*. New York: Wiley, New York.

Fluss, R., Faraggi, D. and Reiser, B. (2005). Estimation of the Youden Index and its associated cutoff point. *Biometrical Journal*, **47**, 458-472.

Gao, S., Hui, S.L., Hall, K.S. and Hendrie, H.C. (2000). Estimating disease prevalence from two-phase surveys with non-response at the second phase. *Statistics in Medicine*, **19**, 2101-2114.

Garrett, E.S., Eaton, E.E. and Zeger, S. (2002). Methods for evaluating the performance of diagnostic tests in the absence of a gold standard: a latent class model approach. *Statistics in Medicine*, **21**, 1289-1307.

Gini, C. (1912). Variabilit'a e mutabilit'a. Reprinted in *Memorie di Metodologica Statistica*, (Ed. Pizetti E. and Salvemini, T.), 1955, Libreria Eredi Virgilio Veschi, Rome.

Girling, A.J. (2000). Rank statistics expressible as integrals under P-P-plots and receiver operating characteristic curves. *Journal of the Royal Statistical Society, Series B*, **62**, 367-382.

Goddard, M.J. and Hinberg, I. (1990). Receiver operating characteristic (ROC) curves and non-normal data: an empirical study. *Statistics in Medicine*, **9**, 325-337.

Gold, M.R., Siegel, J.E., Russel, L.B. and Weinstein, M.C. (1977). *Cost Effectiveness in Health and Medicine*. Wiley, New York.

Goodenough, D.J., Rossman, K. and Lusted, L.B. (1974). Radiographic applications of receiver operating characteristic (ROC) analysis. *Radiology*, **110**, 89-95.

Goovaerts, P., Jacquez, G.M. and Marcus, A. (2005). Geostatistical and local cluster analysis of high resolution hyperspectral imagery for the detection of anomalies. *Remote Sensing of Environment*, **95**, 351-367.

Gorsevski, P.V., Gessler, P. and Foltz, R.B. (2000). Spatial prediction of landslide hazard using logistic regression and GIS. In: *4th International Conference on Integrating GIS and Environmental Modeling (GIS/EM4), Problems, Prospects and Research Needs*, 110. Banff, Alberta.

Gotway, C.A. and Stroup, W.W. (1997). A generalized linear model approach to spatial data analysis and prediction. *Journal of Agricultural, Biological, and Environmental Statistics*, **2**, 157-178.

Green, D.M. and Swets, J.A. (1966). *Signal Detection Theory and Psychophysics*, Wiley, New York.

Greenhouse, S. and Mantel, N. (1950). The evaluation of diagnostic tests. *Biometrics*, **6**, 399-412.

Guido, R., Schiffman, M., Solomon, D. and Burke, L. (2003). Postcolposcopy management strategies for women referred with low-grade squamous intraepithelial lesions or human papillomavirus DNA-positive atypical squamous cells of undetermined significance: a two-year prospective study. *American Journal of Obstetrics and Gynecology*, **188**, 1401-1405.

Gur, D., King, J.L., Rockette, H.E., Britton, C.A., Thaete, F.L. and Hoy, R.J. (1990). Practical issues of experimental ROC analysis: selection of controls. *Investigative Radiology*, **25**, 583-586.

Guttman, I., Johnson, R.A., Bhattacharyya, G.K. and Reiser, B. (1988). Confidence limits for stress-strength models with explanatory variables. *Technometrics*, **30**, 161-168.

Hajian-Tilaki, K.O., Hanley, J.A., Joseph, L. and Collet, J. (1997). A comparison of parametric and nonparametric approaches to ROC analysis of quantitative diagnostic tests. *Medical Decision Making*, **17**, 94-102.

Han, A.K. (1987). Non-parametric analysis of a generalized regression model. *Journal of Econometrics*, **35**, 303-316.

Hand, D.J. (1986). Recent advances in error rate estimation, *Pattern Recognition Letters*, **4**, 335-46

Hand, D.J. (1997). *Construction and Assessment of Classification Rules.* Wiley, Chichester.

Hand, D.J. (2001). Measuring diagnostic accuracy of statistical prediction rules. *Statistica Neerlandica*, **55**, 3-16.

Hand, D.J. (2005). Good practice in retail credit scorecard assessment. *Journal of the Operational Research Society*, **56**, 1109-1117.

Hand D.J. and Till R.J. (2001). A simple generalisation of the area under the ROC curve for multiple class classification problems. *Machine Learning*, **45**, 171-186.

Hand, D.J., Oliver, J.J. and Lunn, A.D. (1998). Discriminant analysis when the classes arise from a continuum. *Pattern Recognition*, **31**,

641-650.

Hanley, J.A. (1989) Receiver operating characteristic (ROC) methodology: the state of the art. *Critical Reviews in Diagnostic Imaging*, **29**, 307-335.

Hanley, J.A. (1996). The use of the 'binormal' model for parametric ROC analysis of quantitative diagnostic tests. *Statistics in Medicine*, **15**, 1575-1585.

Hanley, J.A. and Hajian-Tilaki, K.O. (1997). Sampling variability of nonparametric estimates of the areas under receiver operating characteristic curves: an update. *Academic Radiology*, **4**, 49-58.

Hanley, J.A. and McNeil, B.J. (1982). The meaning and use of the area under an ROC curve. *Radiology*, **143**, 29-36.

Hanley, J.A. and McNeil, B. (1983). A method of comparing the area under two ROC curves derived from the same cases. *Radiology*, **148**, 839-43.

Hardy, R.J. and Thompson, S.G. (1996). A likelihood approach to meta-analysis with random effects. *Statistics in Medicine*, **15**, 619-629.

Hastie, T., Tibshirani, R. and Friedman, J. (2001). *The Elements of Statistical Learning*. Springer, New York.

Hayes, S.C. (2002). Early intervention or early incarceration? Using a screening test for intellectual disability in the criminal justice system. *Journal of Applied Research in Intellectual Disabilities*, **15**, 120-128.

He X., Metz C.E., Tsui B.M.W., Links J.M., and Frey E.C. (2006). Three-class ROC analysis—a decision theoretic approach under the ideal observer framework. *IEEE Transactions on Medical Imaging*, **25**, 571-581.

Heagerty, P.J. and Pepe, M.S. (1999). Semiparametric estimation of regression quantiles with application to standardizing weight for height and age in US children. *Applied Statistics*, **48**, 533-551.

Hellmich, M., Abrams, K.R., Jones, D.R. and Lambert, P.C. (1998). A Bayesian approach to a general regression model for ROC curves. *Medical Decision Making*, **18**, 436-443.

Hellmich, M., Abrams, K.R. and Sutton, A.J. (1999). Bayesian approaches to meta-analysis of ROC curves. *Medical Decision Making*, **19**, 252-264.

Henkelman, R.M., Kay, I. and Bronskill, M.J. (1990). Receiver Operating Characteristic (ROC) analysis without truth. *Medical Decision Making*, **10**, 24-29.

Holliday, J.R., Rundle, J.B., Turcotte, D.L., Klein, W., Tiampo, K.F. and Donellan, A. (2006). Space-time clustering and correlations of major earthquakes. *Physical Review Letters*, **97**, 238501.

Hsieh, F. and Turnbull, B.W. (1996). Nonparametric and semiparametric estimation of the receiver characteristic curve. *Annals of Statistics*, **24**, 25-40.

Irwig, L., Macaskill, P., Glasziou, P. and Fahey, M. (1995). Meta-analytic methods for diagnostic test accuracy. *Journal of Clinical Epidemiology*, **48**, 119-130.

Ishwaran, H. and Gatsonis, C.A. (2000). A general class of hierarchical ordinal regression models with applications to correlated ROC analysis. *The Canadian Journal of Statistics*, **28**, 731-750.

Ishwaran, H. and James, L.F. (2002). Approximate Dirichlet process computing in finite normal mixtures: smoothing and prior information. *Journal of Computational and Graphical Statistics*, **11**, 508-532.

Jamain, A. (2004). *Meta-Analysis of Classification Methods*. Unpublished PhD thesis, Department of Mathematics, Imperial College, London.

Janes, H. and Pepe, M.S. (2006). Adjusting for covariate effects on classification accuracy using the covariate-adjusted ROC curve. Technical Report 283, UW Biostatistics Working Paper Series. Available at http://www.bepress.com/uwbiostat/paper283

Janes, H. and Pepe, M.S. (2007). Adjusting for covariates in studies of diagnostic, screening, or prognostic markers: an old concept in a new setting. Technical Report 310, UW Biostatistics Working Paper Series. Available at http://www.bepress.com/uwbiostat/paper310

Janes, H. and Pepe, M.S. (2008). Matching in studies of classification accuracy: implications for analysis, efficiency, and assessment of incremental value. *Biometrics*, **64**, 1-9.

Janes, H., Longton, G.M. and Pepe, M.S. (2008). Accommodating covariates in ROC analysis. Technical Report 322, UW Biostatistics Working Paper Series. Available at http://www.bepress.com/uwbiostat/paper322

Kessler, R.C. (2002). The categorical versus dimensional assessment controversy in the sociology of mental illness. *Journal of Health and Social Behavior*, **43**, 171-188.

Krzanowski, W.J. and Marriott, F.H.C. (1995). *Multivariate Analysis Part 2: Classification, Covariance Structures and Repeated Measurements*. Arnold, London.

Kucharski, L.T., Toomey, J.P. and Fila, K. (2007). Detection of malingering of psychiatric disorder with the Personality Assessment Inventory: an investigation of criminal defendants. *Journal of Personality Assessment*, **88**, 25-32.

Kuncheva, L.I. (2004). *Combining Pattern Classifiers, Methods and Algorithms*, Wiley, New Jersey.

Lachenbruch, P.A. (1977). Covariance adjusted discriminant functions. Annals of the Institute of Statistical Mathematics, **29**, 247-257.

Landgrebe, T.C.W. and Duin, R.P.W. (2007). Approximating the multiclass ROC by pairwise analysis. *Pattern Recognition Letters*, **28**, 1747-1758.

Landgrebe, T.C.W. and Duin, R.P.W. (2008). Multiclass ROC approximation by decomposition via confusion matrix perturbation analysis. *IEEE Transactions on Pattern Analysis and Machine Intelligence*, **30**, 810-822.

Le, C.T. (2006). A solution for the most basic optimization problem associated with an ROC curve. *Statistical Methods for Medical Research*, **15**, 571-584.

Lee, P. (2004). *Bayesian Statistics: An Introduction* (3rd Edition). Arnold, London.

Lee, W.C. (1999). Probabilistic analysis of global performances of diagnostic tests: interpreting the Lorenz curve-based summary measures. *Statistics in Medicine*, **18**, 455-471.

Lee, W.C. and Hsiao, C.K. (1996). Alternative summary indices for the receiver operating characteristic curve. *Epidemiology*, **7**, 605-611.

Lin, J.-H. and Haug, P.J. (2008). Exploiting missing clinical data in Bayesian network modeling for predicting medical problems. *Journal of Biomedical Informatics*, **41**, 1-14.

Little, R.J.A. and Rubin, D.B. (2002). *Statistical Analysis with Missing Data* (2nd Edition). Wiley, New York.

Liu, A., Schisterman, E.F. and Wu, C. (2005). Nonparametric estimation and hypothesis testing on the partial area under receiver characteristic curves. *Communications in Statistics - Theory and Methods*, **34**, 2077-2088.

Lloyd, C.J. (1998). The use of smoothed ROC curves to summarise and compare diagnostic systems. *Journal of the American Statistical Association*, **93**, 1356-1364.

Lloyd, C.D. (2002). Semiparametric estimation of ROC curves based on binomial regression modelling. *Australian and New Zealand Journal of Statistics*, **44**, 75-86.

Lorenz, M.O. (1905). Methods of measuring the concentration of wealth. *American Statistical Association*, **9**, 209-219.

Lusted, L.B. (1968). *Introduction to Medical Decision Making*. Thomas, Springfield.

Ma, G. and Hall, W.J. (1993). Confidence bands for receiver operating characteristic curves. *Medical Decision Making*, **13**, 191-197.

Ma, S. and Huang, J. (2005). Regularized ROC method for disease classification and biomarker selection with microarray data. *Bioinformatics*, **21**, 4356-4362.

Ma, S., Song, X. and Huang, J. (2006). Regularized binormal ROC method in disease classification with microarray data. *BMC Bioinformatics*, **7**, 253.

Mamitsuka, H. (2006). Selecting features in microarray classification using ROC curves. *Pattern Recognition*, **39**, 2393-2404.

Marascuilo, L.A. and McSweeney, M. (1977). *Nonparametric and Distribution-free Methods for the Social Sciences*. Brooks-Cole Publishing Company, Monterey, California.

Mason, S.J. and Graham, N.E. (2002). Areas beneath the relative operating characterists (ROC) and relative operating levels (ROL) curves: statistical significance and interpretation. *Quarterly Journal of the Meteorological Society*, **128**, 2145-2166.

McClish, D.K. (1989). Analyzing a portion of the ROC curve. *Medical Decision Making*, **9**, 190-195.

McClish, D.K. (1990). Determining the range of false-positives for which ROC curves differ. *Medical Decision Making*, **10**, 283-287.

McCullagh, P. and Nelder, J.A. (1989). *Generalized Linear Models* (2nd Edition). Chapman & Hall, New York.

REFERENCES

McIntosh, M.W. and Pepe, M.S. (2002). Combining several screening tests: optimality of the risk score. *Biometrics*, **58**, 657-664.

McLachlan, G.J. (1992). *Discriminant Analysis and Pattern Recognition*. Wiley, New York.

Mee, R.W. (1990). Confidence intervals for probabilities and tolerance regions based on a generalization of the Mann-Whitney statistic. *Journal of the American Statisticsl Association*, **85**, 793-800.

Merz, C.J. and Murphy, P.M. (1996). UCI depository of machine learning database. http://www.ic.uci.edu/ mlearn/MLRepository.html

Metz, C.E. (1978) Basic principles of ROC analysis. *Seminars in Nuclear Medicine*, **8**, 283-98.

Metz, C.E. (1989). Some practical issues of experimental design and data analysis in radiological ROC studies. *Investigative Radiology*, **24**, 234-245.

Metz, C.E. (2008). ROC analysis in medical imaging: a tutorial review of the literature. *Radiological Physics & Technology*, **1**, 2-12.

Metz, C. and Kronman, H. (1980). Statistical significance tests for binormal ROC curves. *Journal of Mathematical Psychology*, **22**, 218-243.

Metz, C., Herman, B.A., and Roe, C.A. (1998). Statistical comparison of two ROC estimates obtained from partially paired classes. *Medical Decision Making*, **18**, 110-121.

Metz, C.E., Herman, B.A. and Shen, J. (1998). Maximum likelihood estimation of receiver operating characteristic (ROC) curves from continuously distributed data. *Statistics in Medicine*, **17**, 1033-1053.

Metz, C., Wang, P., and Kronman, H. (1984). A new approach for testing the significant differences between ROC curves measured from correlated data. In F. Deconinck (ed.) *Information Processing in Medical Imaging*. Nijihoff, The Hague, The Netherlands.

Mossman, D. (1999). Three-way ROCs. *Medical Decision Making*, **19**, 78-89.

Müller, P., Erkanli, A. and West, M. (1996). Bayesian curve fitting using multivariate normal mixtures. *Biometrika*, **83**, 67-79.

Nakas, C.T. and Yiannoutsos, C.T. (2004). Ordered multiple-class ROC analysis with continuous measurements. *Statistics in Medicine*, **23**, 3437-3449.

Neyman, J. and Pearson, E.S. (1933). On the problem of the most efficient tests of statistical hypotheses. *Philosophical Transactions of the Royal Society, London*, **A231**, 289-337.

Nielsen, S.S., Gronbak, C., Agger, J.F. and Houe, H. (2002). Maximum-likelihood estimation of sensitivity and specificity of ELISAs and faecal culture for diagnosis of paratuberculosis. *Preventitive Veterinary Medicine*, **53**, 191-204.

Nock, R. (2003). Complexity in the case against accuracy estimation. *Theoretical Computer Science*, **301**, 143-165.

Normand, S-L.T. (1999). Tutorial in biostatistics. Meta-analysis: formulating, evaluating, combining and reporting. *Statistics in Medicine*, **18**, 321-359.

Obuchowski, N.A. (1998). Sample size calculations in studies of test accuracy. *Statistical Methods in Medical Research*, **7**, 371-392.

Obuchowski, N.A. (2003). Receiver operating characteristic curves and their use in radiology. *Radiology*, **229**, 3-8.

Obuchowski, N.A. (2004). How many observers are needed in clinical studies of medical imaging? *American Journal of Radiology*, **182**, 867-869.

Obuchowski, N.A. (2005). ROC analysis. *American Journal of Radiology*, **184**, 364-372.

Obuchowski, N.A. and McClish, D.K. (1997). Sample size determination for diagnostic accuracy studies involving binormal ROC curve indices. *Statistics in Medicine*, **16**, 1529-1542.

Obuchowski, N.A. and Rockette, H.E. (1995). Hypothesis testing of the diagnostic accuracy for multiple diagnostic tests: an ANOVA approach with dependent observations. *Communications in Statistics, Simulation & Computation*, **24**, 285-308.

Obuchowski, N.A., Beiden, S.V., Berbaum, K.S., Hillis, S.L., Ishwaran, H., Song, H.H. and Wagner, R.F. (2004). Multireader, multicase receiver operating characteristic analysis: an empirical comparison of five methods. *Academic Radiology*, **11**, 980-995.

Obuchowski, N.A., Lieber, M.L. and Wians, F.H. (2004). ROC curves in clinical chemistry: uses, misuses, and possible solutions. *Clinical Chemistry*, **50**, 1118-1125.

Ogilvie, J.C. and Creelman, C.D. (1968). Maximum likelihood estimation of ROC curve parameters. *Journal of Mathematical Psychology*, **5**, 377-391.

Owen, D.B., Craswell, K.J. and Hanson, D.L. (1964). Non-parametric upper confidence bounds for $P(Y < X)$ and confidence limits for $P(Y < X)$ when X and Y are normal. *Journal of the American Statisticsl Association*, **59**, 906-924.

Peng, F.C. and Hall, W.J. (1996). Bayesian analysis of ROC curves using Markov-chain Monte Carlo methods. *Medical Decision Making*, **16**, 404-411.

Peng, L. and Zhou, X.-H. (2004). Local linear smoothing of receiver operating characteristic (ROC) curves. *Journal of Statistical Planning and Inference*, **118**, 129-143.

Pepe, M.S. (1997). A regression modelling framework for receiver operating characteristic curves in medical diagnostic testing. *Biometrika*, **84**, 595-608.

Pepe, M.S. (1998). Three approaches to regression analysis of receiver operating characteristic curves for continuous test results. *Biometrics*, **54**, 124-135.

Pepe, M.S. (2000). An interpretation for the ROC curve and inference using GLM procedures. *Biometrics*, **56**, 352-359.

Pepe, M.S. (2003). *The Statistical Evaluation of Medical Tests for Classification and Prediction*. University Press, Oxford.

Pepe, M.S. and Cai, T. (2004). The analysis of placement values for evaluating discriminatory measures. *Biometrics*, **60**, 528-535.

Pepe, M.S., Cai, T. and Longton, G. (2006). Combining predictors for classification using the area under the receiver operating characteristic curve. *Biometrics*, **62**, 221-229.

Perkins, N.J. and Schisterman, E.F. (2006). The inconsistency of "optimal" cutpoints obtained using two criteria based on the receiver operating characteristic curve. *American Journal of Epidemiology*, **163**, 670-675.

Peterson, W.W., Birdsall, T.G. and Fox, W.C. (1954). The theory of signal detectability. *Transactions of the IRE Professional Group on Information Theory, PGIT*, **2-4**, 171-212.

Pfeiffer, R.M. and Castle, P.E. (2005). With or without a gold standard. *Epidemiology*, **16**, 595-597.

Provost F. and Fawcett T. (1997). Analysis and visualisation of classifier performance: comparison under imprecise class and cost distributions. In: *Proceedings of the Third International Conference on Knowledge Discovery and Data Mining*, AAAI Press, Menlo Park, CA, 43-48.

Provost F. and Fawcett T. (1998). Robust classification systems for imprecise environments. In: *Proceedings of the 15th National Conference on Artificial Intelligence*, AAAI Press, Madison, WI, 706-707.

Provost, F., Fawcett, T. and Kohavi, R. (1998). The case against accuracy estimation for comparing induction algorithms. In: *Proceedings of the 15th International Conference on Machine Learning*, pp 445-453, July 24-27 1998.

Qin, G. and Zhou, X.-H. (2006). Empirical likelihood inference for the area under the ROC curve. *Biometrics*, **62**, 613-622.

Qin, J. and Zhang, B. (2003). Using logistic regression procedures for estimating receiver operating characteristic curves. *Biometrika*, **90**, 585-596.

Ransohoff, D.F. and Feinstein, A.R. (1978). Problems of spectrum and bias in evaluating the efficacy of diagnostic tests. *New England Journal of Medicine*, **299**, 926-930.

Rao, C.R. (1949). On some problems arising out of discrimination with multiple characters. *Sankhya*, **9**, 343-366.

Reiser, B. (2000). Measuring the effectiveness of diagnostic markers in the presence of measurement error through the use of ROC curves. *Statistics in Medicine*, **19**, 2115-2129.

Reiser, B. and Faraggi, D. (1994). Confidence bounds for $\Pr(a'X > b'Y)$. *Statistics*, **25**, 107-111.

Reiser, B. and Guttman, I. (1986). Statistical inference for $P(Y < X)$: the normal case. *Technometrics*, **28**, 253-257.

Ren, H., Zhou, X.-H. and Liang, H. (2004). A flexible method for estimating the ROC curve. *Journal of Applied Statistics*, **31**, 773-784.

Ripley, B.D. (1996) *Pattern Recognition and Neural Networks*. Cambridge University Press, Cambridge.

Rotnitzky, A., Faraggi, D. and Schisterman, E. (2006). Doubly robust estimation of the area under the receiver-operating characteristic curve in the presence of verification bias. *Journal of the American Statistical Association*, **101**, 1276-1288.

Schafer, J.L. (1999). Multiple imputation: a primer. *Statistical Methods in Medical Research*, **8**, 3-15.

Schiavo, R. and Hand, D.J. (2000). Ten more years of error rate research. *International Statistical Review*, **68**, 295-310.

Schisterman, E.F., Faraggi, D., Reiser, B. and Trevisan, M. (2001). Statistical inference for the area under the receiver operating characteristic curve in the presence of random measurement error. *American Journal of Epidemiology*, **154**, 174-179.

Schisterman, E.F., Faraggi, D. and Reiser, B. (2004). Adjusting the generalized ROC curve for covariates. *Statistics in Medicine*, **23**, 3319-3331.

Scott M.J.J., Niranjan M. and Prager R.W. (1998). Parcel: feature subset selection in variable cost domains. Technical Report CUED/F-INFENG/TR.323, Cambridge University Engineering Department.

Sen, P.K. (1960). On some convergence properties of U-statistics. *Calcutta Statistical Association Bulletin*, **10**, 1-18.

Shewhart, W. (1931). *Economic Control of Quality of Manufactured Products*. van Norstand, New York.

Shieh, G., Show-Li, J. and Randles, R.H. (2006). On power and sample size determination for the Wilcoxon-Mann-Whitney test. *Nonparametric Statistics*, **18**, 33-43.

Silverman, B.W. (1986). *Density Estimation in Statistics and Data Analysis*. Chapman & Hall, London.

Simonoff, J.S., Hochberg, Y. and Reiser, B. (1986). Alternative estimation procedures for $P(X < Y)$ in categorized data. *Biometrics*, **42**, 815-907.

Smith, P.J. and Thompson, T.J. (1996). Correcting for confounding in analyzing receiver operating characteristic curves. *Biometrical Journal*, **38**, 857-863.

Smith, T.C., Spiegelhalter, D.J. and Thomas, A. (1995). Bayesian approaches to random effects meta-analysis: a comparative study. *Statistics in Medicine*, **14**, 2685-2699.

Song, H.H. (1997). Analysis of correlated ROC areas in diagnostic testing. *Biometrics*, **53**, 370-382.

Spiegelhalter, D.J., Thomas, A., Best, N.G. and Gilks, W.R. (1995). *BUGS: Bayesian Inference Using Gibbs Sampling, Version 0.50*. Cambridge: MRC Biostatistics Unit.

Sprent, P. (1989). *Applied Nonparametric Statistical Methods*. Chapman & Hall, London.

Spritzler, J., DeGruttola, V.G. and Pei, L. (2008). Two-sample tests of area-under-the-curve in the presence of missing data. *The International Journal of Biostatistics*, **4**, 18 pp., The Berkeley Electronic Press.

Stepanova, M. and Thomas, L.C. (2002). Survival analysis methods for personal loans data. *Operations Research*, **50**, 277-289.

Stephan, C., Wesseling, S., Schink, T. and Jung, K. (2003). Comparison of eight computer programs for receiver-operating characteristic analysis. *Clinical Chemistry*, **49**, 433-439.

Stover, L., Gorga, M.P., Neely, S.T. and Montoya, D. (1996). Toward optimizing the clinical utility of distortion product otoacoustic emission measurements. *Journal of the Acoustical Society of America*, **100**, 956-967.

Swets, J.A. and Pickett, R.M. (1982). *Evaluation of Diagnostic Systems: Methods from Signal Detection Theory*. Academic Press, New York.

Tanner, M. and Wong, W. (1987). The calculation of posterior distributions by data augmentation. *Journal of the American Statistical Association*, **82**, 528-550.

Thomas, L.C., Banasik, J. and Crook, N. (1999). Not if but when loans default. *Journal of the Operational Research Society*, **50**, 1185-1190.

Thompson, M.L. and Zucchini, W. (1989). On the statistical analysis of ROC curves. *Statistics in Medicine*, **8**, 1277-1290.

Thuiller, W., Vayreda, J., Pino, J., Sabate, S., Lavorel, S. and Gracia, C. (2003). Large-scale environmental correlates of forest tree distributions in Catalonia (NE Spain). *Global Ecology & Biogeography*, **12**, 313-325.

Toh, K-A., Kim, J. and Lee, S. (2008). Maximizing area under ROC curve for biometric scores fusion. *Pattern Recognition*, **41**, 3373-3392.

Toledano, A.Y. and Gatsonis, C. (1996). Ordinal regression methodology for ROC curves derived from correlated data. *Statistics in Medicine*, **15**, 1807-1826.

Tosteson, A.N.A. and Begg, C.B. (1988). A general regression methodology for ROC curve estimation. *Medical Decision Making*, **8**, 204-215.

Tosteson, D., Buonaccorsi, J.P., Demidenko, E. and Wells, W.A. (2005). Measurement error and confidence intervals for ROC curves. *Biometrical Journal*, **47**, 409-416.

Van der Heiden, G.J.M.G., Donders, A.R.T., Stijnen, T. and Moons, K.G.M. (2006). Imputation of missing values is superior to complete case analysis and the missing-indicator method in multivariable diagnostic research: a clinical example. *Journal of Clinical Epidemiology*, **59**, 1102-1109.

Venables, W.N. and Ripley, B.D. (2002). *Modern Applied Statistics with S*. 4th Edition. Springer, New York.

Venkatraman, E.S. (2000). A permutation test to compare receiver operating characteristic curves. *Biometrics*, **56**, 1134-1138.

Venkatraman, E.S. and Begg, C.B. (1996). A distribution-free procedure for comparing Receiver Operating Characteristic curves from a paired experiment. *Biometrika*, **83**, 835-848.

Verde, M.F. and Rotello, C.M. (2004). ROC curves show that the revelation effect is not a single phenomenon. *Psychonomic Bulletin & Review*, **11**, 560-566.

Vida, S. (1993). A computer program for non-parametric receiver operating characteristic analysis. *Computer Methods and Programs in Biomedicine*, **40**, 95-101.

Wang, C., Turnbull, B.W., Gröhn, Y.T. and Nielsen, S.S. (2007). Nonparametric estimation of ROC curves based on Bayesian models when the true disease state is unknown. *Journal of Agricultural, Biological and Environmental Statistics*, **12**, 128-146.

Webb, A.R. (2002). *Statistical Pattern Recognition* (2nd Edition). Wiley, Chichester.

Weinstein, M.C., Fineberg, H.V., Elstein, A.S., Frasier, H.S., Neuhauser, D., Neutra, R.R. and McNeil, B.J. (1996). *Clinical Decision Analysis*. W.B. Saunders, Philadelphia.

Wieand, S., Gail, M.H., James, B.R. and James, K.L. (1989). A family of nonparametric statistics for comparing diagnostic markers with paired or unpaired data. *Biometrika*, **76**, 585-592.

Wilks, D.S. (2001). A skill score based on economic value for probability forecasts. *Meteorological Applications*, **8**, 209-219.

Williams, P., Hand, D.J. and Tarnopolsky, A. (1982) The problem of screening for uncommon disorders a comment on the Eating Attitudes Test, *Psychological Medicine*, **12**, 431-4.

Wolfe, D.A. and Hogg, R.V. (1971). On constructing statistics and reporting data. *American Statistician*, **25**, 27-30.

Wolfowitz, J. (1957). The minimum distance method. *Annals of Mathematical Statistics*, **28**, 75-88.

Yan L., Dodier R., Mozer M.C. and Wolniewicz R. (2003). Optimizing classifier performance via the Wilcoxon-Mann-Whitney statistics. In: *Proceedings of the 20th International Conference on Machine Learning, ICML-2003*, eds. T.Fawcett and N.Mishra.

Zheng, Y. and Heagerty, P.J. (2004). Semiparametric estimation of time-dependent ROC curves for longitudinal marker data. *Biostatistics*, **5**, 615-632.

Zhou, X.-H. and Harezlak, J. (2002). Comparison of bandwidth selection methods for kernel smoothing of ROC curves. *Statistics in Medicine*, **21**, 2045-2055.

Zhou, X.-H., Obuchowski, N.A. and McClish, D.K. (2002). *Statistical Methods in Diagnostic Medicine*. Wiley, New York.

Zou, K.H. (2001). Comparison of correlated receiver operating characteristic curves derived from repeated diagnostic test data. *Academic Radiology*, **8**, 225-233.

Zou, K.H. and Hall, W.J. (2000). Two transformation models for estimating an ROC curve derived from continuous data. *Journal of Applied Statistics*, **27**, 621-631.

Zou, K.H., Hall, W.J. and Shapiro, D.E. (1997). Smooth nonparametric receiver operating characteristic (ROC) curves for continuous diagnostic tests. *Statistics in Medicine*, **16**, 2143-2156.

Zweig, M.H. and Campbell, G. (1993). Receiver-operating characteristic (ROC) plots: a fundamental evaluation tool in clinical medicine. *Clinical Chemistry*, **39**, 561-577.

Index

AccuROC software, 202
Adjustments
    AUC, 97-99, 164-165
    curves, 89-97
    direct, 91-95
    indirect, 89-91
    partial AUC, 99-100
    summary statistics, 97-102
    two-sample tests, 164-165
Alarms, 186, 189
Alternatives to ROC curves, 141-145
Analyse-it software, 202
Analysis, see also Monte Carlo analysis
    Bayesian methods, 125-131
    complete case, verification bias, 167
    data sets, software, 202
    design and interpretation issues, 178-179
    medical imaging, 178-179
    meta-analysis, 127-131
    survival, 196-197
Anorexia nervosa, 11
ANOVA, 178
Applications
    atmospheric sciences, 184-187, 189-190
    biosciences, 192-195
    covariates, 101-102
    credit card transactions, 2
    email message filtering, 2
    experimental psychology, 198-199
    finance, 1, 195-197
    fundamentals, 14, 181-182
    gene expression patterns, 2
    geosciences, 190-192
    machine learning, 182-184
    medical diagnosis, 1
    ROC curve, adjustment, 95-97
    sociology, 199-200
    speech recognition systems, 1
Arbitrary cost matrix, 156
Area under the curve (AUC), see also Partial area under the curve (PAUC)
    atmospheric sciences, 185
    Bayesian parametric estimation, 135
    biosciences, 193-195
    classifier performance assessment, 11
    comparison of curves, 113-114
    confidence intervals, 79
    doubly robust estimator, 170
    estimation classifier parameters by maximization, 154
    further reading, 86
    hypothesis testing, 80-82
    interpretation of, 26
    Mamitsuka's approach, 195
    measurement errors, 83-85
    measures for nonbinary outcomes, 154-155
    multiple classes, curves, 150-154
    optimum threshold choice, 175
    parametric curve fitting, 47
    sample size calculations, 78
    separation test, P and N pop-

ulation scores, 77-78
summary indices, 26-28, 65-69
summary statistics adjustment, 97-99
two-sample tests, adjustments, 164-165
Wald statistic, 178-179
Artificial sampling distribution, 104-105
Asymptotic properties
    AUC, 78, 98, 99, 114
    comparing AUCs, two ROC curves, 114
    confidence intervals, 67
    empirical estimator, 58, 60
    issues, 58
    matching in case-control studies, 104
    nonparametric estimation, 62-63
    parametric curve fitting, 61-62
Atmospheric sciences, 184-187, 189-190
Attenuation, 162
AUC, see Area under the curve (AUC)
Average map, 192

Background, 1-2, 14
Back-transformation, 68
Bacterial meningitis markers, 83
Banking industry, 172, see also Credit scoring; Finance
Barycentric coordinate spaces, 151
Baseline predictors, 102-103
Bayesian methods
    empirical Bayes, 130-131
    frequentist methods, 128-130
    full Bayes, 131
    fundamentals, 13, 123-125
    further reading, 139
    general ROC analysis, 125-127
    incremental value, 103
    meta-analysis, 127-131
    missing values, 164
    nonparametric estimation, 137-139
    parametric estimation, 134-136
    uncertain/unknown group labels, 132-139
Bayes' theorem, 10-11
Benchmarks, machine-learning, 183
Berkson model, 134
Bias
    corrected estimates, 168
    estimation, 39, 58
    fundamentals, 165-166
    issues, 157
    measurement errors, 83-84
    missing values, 165
    reject inference, 171-172
    training data and sets, 40
    verification bias, 166-171
Binary regression methods
    adjustment of partial AUC, 99-100
    ROC curves estimation, 56-57
    sampling properties and confidence intervals, 63
Binary variables, 2, 99
Binomial model and form
    AUC, 97-98, 114
    Bayesian methods, 125, 134-135
    curves comparison, 110, 115-116, 121-122
    direct covariate adjustment, 94
    estimation, 12, 66
    further reading, 122
    hypothesis testing, 82
    indirect covariate adjustment, 89
    measurement errors, 84-85
    parametric curve fitting, 49-50, 52-53, 61
    parametric estimation, 134-135
    partial AUC, 69
    population ROC curves, 31-32, 34-35
Biomarkers, 193-195

INDEX

Biometrics, 4
Biosciences, 192-195
Black box routines, 39
Blinding, medical imaging, 177
Blood pressure, 83, 164
Bootstrapping
    asymptotic results, 58
    binary regression methods, 63
    classifier performance assessment, 7
    covariate adjustment, 97
    DPOAE hearing-test data, 101
    Youden Index, 100-101
Box-Cox transformation, 47, 66

Carry-over effects, 176
Casewise deletion, 161
Cattle, 136, 138
cdf, see Cumulative distribution function (cdf)
Censored observations, 196
Central limit theorem, 31
Cerebrospinal fluid, 83
Cervical cancer, 132
Chance diagonal
    AUC test, 77-78
    comparing two ROC curves, 108
    empirical estimation, 43
    fundamentals, 19-20, 30
    population scores, tests of separation, 76
Chi-square distribution, 68
Classes, multiple, 13-14, 147-154
Classification
    covariate adjustment, 88
    fundamentals, 2-6
    further reading, 14-15
    incremental value, 103
    optimal threshold, 70-71
    rule and error rates, 39-41
    scores, 88, 98, 125-126
    trees, 39
Classifier
    parameters, estimation, 154
    performance assessment, 6-11

    successfulness, 19-20
Class priors, 10
Cleveland Clinic (Department of Quantitative Health Sciences), 201
Cochran's statistic, 129
Combinatorial approaches, 69
Commercial software, 202
Comparisons
    AUC, 113-114
    binormal case, 115-116
    curves, 113-122
    fundamentals, 13, 107-109
    further reading, 122
    identifying differences, 121-122
    nonparametric approach, 116-120
    regression approaches, 120-121
    summary statistics, two ROC curves, 109-112
Complementary conditional rates, 9
Complementary marginal probability, 9
Complete case (CC) analysis, 167
Component index, Bayesian methods, 126
Concave function, 146
Conclusive results, 81
Conditional distributions, 94-95
Conditional imputation, 162
Conditional probabilities, 8-10
Confidence intervals
    identifying curve differences, 121-122
    measurement errors, 84-85
    ROC curves estimation, 57-63
    sample size calculations, 79
    summary indices, 67-69
Continuous (numerical) variables, 3
Continuous scores, 23, see also Scores
Convex hull ROC curves, 145-147, 156
Cost/loss ratio, 189

Costs and cost-weighting
  curves, 143-144
  error rate, minimum, 30
  further reading, 74
  misallocation/misclassification, 70-72, 150
  misclassification rates, 24, 108, 112
  optimum threshold choice, 173
  ratio, 71, 74
Covariance, 108, 111-114
Covariates
  applications, 95-97, 101-102
  AUC, 97-99
  curve adjustment, 89-97
  curves comparison, 110
  direct adjustment, 91-95
  fundamentals, 12-13, 87-88
  further reading, 105
  incremental value, 102-104
  indirect adjustment, 89-91
  matching in case-control studies, 104-105
  partial AUC adjustment, 99-100
  summary statistics adjustment, 97-102
  Youden Index, 100-101
Cows, 136, 138
Cox's proportional hazards model, 197
Credit card transactions, 2, see also Finance
Credit scoring, see also Finance
  applications, 1
  fundamentals, 195
  historical developments, 14
  reject inference, 171-172
  ROC curves, multiple classes, 148
Cross-correlations, 99
Cross-validation, 40
Cumulative distribution function (cdf)
  binary regression methods, 57
  binormal model, 32
  nonparametric approach, 116

Dairy cattle, 136, 138
Data-based index, 63
Data mining, 4, 202
Data sets, 5, 202
Degree of belief, 123
Degrees of freedom, 137-138
Deletion of observed values, 161
DeLong estimate of variance, 68
Delta method, 85
Density functions, 54, 56, 126
Descriptive variables, 3
Design, medical imaging, 176-177
Design and interpretation issues
  analysis and interpretation, 178-179
  bias in studies, 165-172
  deletion of observed values, 161
  design, 176-177
  fundamentals, 14, 157-158
  further reading, 179
  implications, 163-165
  imputation of missing values, 162-163
  maximum likelihood, 163
  mechanisms, 158-160
  medical imaging, 176-179
  methodology, 160-163
  missing values, 158-165
  optimum threshold choice, 172-175
  reject inference, 171-172
  verification bias, 166-171
Design sets, classification, 5
Diabetes, 45
Diagnostic markers, 82-83
Diagnostic medicine, 35, 104, 164
Diastolic blood pressure, 83
Differential costs, misclassification, 30
Direct adjustment, 91-95
Dirichlet properties, 137-138, see also Mixed Dirichlet process (MDP)

INDEX

Discrete (numerical) variables, 3
Disease screening, 104
Distortion product otoacoustic emissions (DPOAE) test, 95-96, 101
DNA traces, 4
Dorfman and Alf procedure, 49-50, 52, 62
Doubly robust estimator, AUC, 170
DPOAE, see Distortion product otoacoustic emissions (DPOAE) test
DPOAE hearing-test data, 101
Duchenne muscular dystrophy, 175

Earthquakes, see Geoscience applications
Earth sciences, see Geoscience applications
ECMRWF, see European Center for Medium Range Weather Forecasts (ECMRWF)
Electronic signal detection, 74
ELISA, see Enzyme-linked immunosorbent assay (ELISA)
Email message filtering, 2
EM algorithm, 133, 163
Empirical approach
    likelihood approach, 68
    optimum threshold choice, 174
Empirical estimator
    AUC confidence intervals, 67
    curves estimation, 41-43, 45, 47
    sampling properties and confidence intervals, 58-60
Entire curves comparison
    binormal case, 115-116
    fundamentals, 115
    nonparametric approach, 116-120
    regression approaches, 120-121
Enzyme-linked immunosorbent assay (ELISA), 136, 139

Error rates, see Misclassification and error rates
Estimation, see also Inference on single curves
    AUC, 65-69
    binary regression methods, 56-57, 63
    classification rule and error rates, 39-41
    confidence intervals, 57-63, 67-69
    empirical estimator, 41-43, 45, 47, 58-60
    estimation, 65-67
    fundamentals, 12, 37-39
    further reading, 74
    LC index, 71-72, 74
    loss difference plots, 71-72, 74
    nonparametric estimation, 53-54, 56, 62-63
    optimal classification threshold, 70-71
    parametric curve fitting, 47, 49-50, 52-53, 61-62
    partial AUC, 69-70
    sampling properties, 57-63
    summary indices, 63, 65-74
Estimation classifier parameters, 154
European Center for Medium Range Weather Forecasts (ECMRWF), 184, 185
Experimental psychology, 198-199
Extra-domain error rates, 191
Extra sum of squares, 103

Faces, classifying, 4
False positive/negative rates
    classifier performance assessment, 8-10
    convex hull ROC curves, 145-146
Fecal culture binary test, 139
FI, see Full imputation (FI)
Finance, 83, 195-197, see also Credit scoring

Fingerprints, 4
Fisher's Information, 171
Fixed-effects framework, 128-129
Fluctuation map, 192
Forecasting systems, see Atmospheric sciences
Fred Hutchinson Cancer Center (University of Washington), 201
Frequentist methods, 13, 128-130
Full imputation (FI), 168
Function S, 4-5, 17
Further reading, see also References
    arbitrary cost matrix, 156
    AUC, 86
    binormal model and form, 122
    classification, 14-15
    convex hull, 156
    covariates, 105
    design and interpretation issues, 179
    electronic signal detection, 74
    estimation, 74
    gold standard, 139
    goodness of fit, 105
    inference on single curves, 86
    Mann-Whitney U-statistic, 86, 105
    Markov-chain Monte Carlo methods, 139
    medical context, 74
    multiple classes, 156
    objects on a single continuum, 156
    paired curves, 122
    partially paired data, 122
    population ROC curves, 35
    psychology, 74
    radiology, 179
    risk score, 105
    ROC curve comparisons, 122
    test of equality, 122
    variants of ROC curves, 155-156
    volume under the surface, 156
Future performance, 7

Gait, 4
Gene expression patterns, 2
Generalized least squares, 62
Generalized linear model (GLM)
    AUC adjustment, 99
    direct covariate adjustment, 92-95
    regression approaches, 120-121
Geoscience applications, 190-192
Gibbs sampling algorithm
    Bayesian methods, 124, 126-127
    nonparametric estimation, 138
    parametric estimation, 136
    uncertain/unknown group labels, 134
Gini coefficient, 27-28, 153
Ginzburg criterion, 192
GLM, see Generalized linear model (GLM)
Gold standard test/diagnosis
    Bayesian parametric estimation, 134-137
    further reading, 139
    issues, 157-158
    uncertain/unknown group labels, 132-133
    verification bias, 166-167
Goodness-of-fit, 49, 105

Hayes Ability Screening Test, 200
Hazard function, survival analysis, 197
Hearing-test data, 95-96, 101
Heart disease, 45, 47
Hierarchical ordinal regression approach, 179
High spatial resolution hyperspectral (HSRH) imagery, 192
Historical background, 1-2, 14
HSRH, see High spatial resolution hyperspectral (HSRH) imagery
Hungarian Institute of Cardiology (Budapest), 45

INDEX

Hyperparameters, 129, 131
Hypersurface, 148-152
Hypertension markers, 83
Hypothesis testing
    AUC test, 78
    binormal case, 115
    curves comparison, 110
    Neyman-Pearson Lemma, 25
    nonparametric approach, 118
    sample size calculations, 78, 80-82

Identical revelation, 198
Identifying curve differences, 121-122
Ignorable missing values, see Missing at random (MAR); Missing completely at random (MCAR)
Imputation of missing values
    full, 168
    inverse probability weighting, 169
    mean score imputation, 168-169
    missing values, 162-163
    semiparametric efficient, 169
Income distributions, 27-28
Incremental value, covariates, 102-104
Indirect adjustment, 89-91
Individuals, see Biometrics; Populations
Induction algorithms, 183
Inference on single curves, see also Estimation
    AUC=0.5 test, 77-78
    confidence intervals, 79
    fundamentals, 12, 75-76
    further reading, 86
    hypothesis tests, 80-82
    Kolmogorov-Smirnov test, 76-77
    measurement errors, 82-85
    sample size calculations, 78-82

    tests of separation, 76-78
Intellectual disability, 200
Intercept, binormal ROC curve, 32
Interpretation, 178-179, see also Design and interpretation issues
Intra-domain error rates, 191
Inverse logit transform, 62-63
Inverse of the hessian matrix, 61
Inverse probability weighting (IPW), 169
Inverse transform, 62
IPW, see Inverse probability weighting (IPW)
Iris patterns, 4
ISI Web of Knowledge, 14
Issues. variants to ROC curves, 154-155

Jackknifing
    asymptotic results, 58
    imputation of missing values, 170
    pseudovalues response, 178
Jaundice, 37
Johne's disease, 136, 138
Joint distribution, 61, 131
Joint effect, 103
Joint likelihood, 56-57
Joint probabilities, 8, 10

Kaufman Brief Intelligence Test, 200
Kernel density methods
    AUC estimation, 66
    empirical estimator, 60
    nonparametric estimation, 54, 56
Kernel estimates, confidence intervals, 68
Kernel method, optimum threshold choice, 175
Kolmogorov-Smirnov statistic and test, 30, 76-77

LABROC software, 50, 54

Landslides, see Geoscience applications
LC, see Loss comparison (LC) index
Least squares regression, 90, 92
Leave-one-out method, 7, 41
Lift curve, 142
Likelihood ratio, 24-25
Linear discriminators, 39, 47
Listwise deletion, 161
Location-scale model, 91
Logistical discrimination, 39
Logistic model, 53
Logit link function and transformation
    AUC adjustment, 98-99
    DPOAE hearing-test data, 101
    incremental value, 104
    nonparametric estimation, 62-63
    partial AUC adjustment, 100
Log-likelihood, 61
Lorenz curve, 27-28, 142
Loss comparison (LC) index
    alternatives to ROC curve, 144
    fundamentals, 30
    further reading, 74
    summary indices, 71-72, 74
Loss difference plots, 30, 71-72, 74

Machine learning
    applications, 182-184
    classification, 4
    software, 202
Mamitsuka's approach, 195
Mann-Whitney U-statistic
    atmospheric sciences, 187
    AUC, 12, 65-67, 99
    comparing curves, 113
    confidence intervals, 67
    estimation, 65-67
    further reading, 86, 105
    imputation of missing values, 170
    machine learning, 184

optimum threshold choice, 175
partial AUC, 70
MAR, see Missing at random (MAR)
Marginal effect, 103
Marginal probabilities, 8-10
Markov blanket filtering, 195
Markov-chain Monte Carlo methods, see also Monte Carlo analysis
    Bayesian methods, 124, 127
    full Bayes, 131
    further reading, 139
    nonparametric estimation, 138
Matched studies, 104-105, 176-177
MATLAB® software, 138, 202
Maximum likelihood (ML)
    AUC, 66-67
    Bayesian nonparametric estimation, 139
    binary regression methods, 57
    confidence intervals, 67
    covariate adjustment, 96
    estimation, 66-67
    frequentist methods, 129-130
    missing values, 163
    parametric curve fitting, 47, 61
    uncertain/unknown group labels, 133
Maximum rank correlation (MRC) estimator, 194
Maximum semiparametric estimators, 57
Maximum vertical distance (MVD)
    AUC test, 77-78
    curves comparison, 108
    fundamentals, 30
    Kolmogorov-Smirnov test, 76-77
    summary indices, 30
MCAR, see Missing completely at random (MCAR)
McNemar's test, 110
MDP, see Mixed Dirichlet process (MDP)

INDEX

Mean score imputation (MSI), 168-169
Mean squared error, 39, 67
Measurements, see also Area under the curve (AUC)
    continuous scores, 23
    errors, inference on single curves, 82-85
    global measure of separability, 11
    multivariate context, 3
    nonbinary outcomes, 154-155
MedCalc software, 201-202
Medical context and applications
    Bayesian methods, 125
    classification, 4, 6
    further reading, 35, 74
    matching in case-control studies, 104
    measurement errors, 83
    multivariate context, 3
    ROC curves, multiple classes, 148
    software, 201-202
Medical imaging, 176-179
Meningitis markers, 83
Meta-analysis, 8, 127-131, see also Analysis
Meteorology, see Atmospheric sciences
Method of moments (MM), 129
Metropolis-Hastings algorithm, 124
Midpoints, AUC estimation, 65
Minimum cost-weighted error rate, 30
Minimum distance estimator, 52
Misclassification and error rates
    alternatives to ROC curve, 143-144
    AUC interpretation, 26
    classifier performance assessment, 8
    cost-weighted, 24, 108, 112
    curves, 12, 150
    differential costs, 30
    further reading, 15
    loss difference plots, 71
    multiple classes, 150
    nonparametric approach, 117-118
    optimal classification threshold, 70-72
    optimum threshold choice, 173
    severity of losses, 30
Missing at random (MAR)
    defined, 159
    ignorable missing value, 160
    imputation of missing values, 169-170
    semiparametric test, 165
    verification bias, 167
Missing completely at random (MCAR)
    defined, 159
    ignorable missing value, 160
    loss of power, 161
    unbiased, 165
    verification bias, 168
Missing-indicator method, 162
Missing not at random (MNAR)
    defined, 159-160
    imputation of missing values, 169-170
    nonignorable mechanism, 160
    verification bias, 167
Missing values
    deletion of observed values, 161
    implications, 163-165
    imputation of missing values, 162-163
    maximum likelihood, 163
    mechanisms, 158-160
    methodology, 160-163
Mixed Dirichlet process (MDP), 126, see also Dirichlet properties
Mixed-effects ANOVA, 178
MM, see Method of moments (MM)
MNAR, see Missing not at random (MNAR)
Model Output Statistics, 189

Moments, see Method of moments (MM)
Monotonicity guarantee, 137-138
Monotonic transformations
 binormal model, 34-35
 cost-weighted misclassification rates, 24
 empirical estimation, 45
 estimation, 56
 nonparametric methods, 56, 116
 parametric curve fitting, 49
 regression approaches, 120
 ROC properties, 20
Monte Carlo analysis, see also Analysis; Markov-chain Monte Carlo methods
 Bayesian methods, 127
 measurement errors, 84, 85
 nonparametric estimation, 63
 uncertain/unknown group labels, 133
MRC, see Maximum rank correlation (MRC) estimator
MRMC, see Multiple reader, multiple case (MRMC) studies
MSI, see Mean score imputation (MSI)
Multiple classes, 13-14, 147-156
Multiple curves, comparing, 108
Multiple imputation, 162-163
Multiple reader, multiple case (MRMC) studies, 177, 178
Multivariable integration, 124
Multivariate variables, classification, 3
Muscular dystrophy, 175
MVD, see Maximum vertical distance (MVD)

Naive Bayes classifier, 45, see also Bayesian methods
National Centers for Environmental Prediction (NCEP), 184
Natural hazards, see Geoscience applications

NCEP, see National Centers for Environmental Prediction (NCEP)
Negative Impression Management scale, 200
Negative predictive value (npv), 9
Nested cross-validation, 41
Neural networks
 classifier type, 39
 curves that cross, 108
 empirical estimation, 45
 loss difference plots, 72, 74
Neyman-Pearson Lemma, 24-25, 103
Noise, biomarkers, 194
Nominal variables, 2
Nonbiased estimator, 65-66
Nonbinary outcomes, 154-155
Nonignorable missing values, see Missing not at random (MNAR)
Noninformative priors, 139
Nonparametric approach
 AUC estimation, 67
 covariate adjustment, 96
 curves, 53-54, 56
 entire curves comparison, 116-120
 estimation, 38, 53-54, 56, 108
 Kolmogorov-Smirnov test, 77
 measurement errors, 84
 partial AUC, 69-70
 sampling properties and confidence intervals, 62-63
 uncertain/unknown group labels, 137-139
Nonstandard object classes, 5-6
N population, see Populations
npv, see Negative predictive value (npv)

Objects on a single continuum, 156
Offender groups, reconviction, 200
Optimal classification threshold, 70-71, 74
Optimum threshold choice, 172-175
Ordered multinomial form, 137-138
Ordinal regression approach, 179

INDEX

Ordinal-score approach, 88, 90
Ordinal variables, 2
Orientation, 20
Osteoporosis, 3

PAI, see Personality Assessment Inventory (PAI)
Paired data
    comparison, 110-112
    curves, 108, 110-112
    further reading, 122
    nonparametric approach, 118-119
Paired-patient, paired-reader, 177
Pairwise deletion, 161
Pancreatic cancer, 96-97
Parametric approach
    distribution-free, 53
    estimation, 38
    measurement errors, 84
    models, AUC estimation, 66
    partial AUC, 69
    ROC curves, 108
    uncertain/unknown group labels, 134-136
Parametric curve fitting
    curves estimation, 47, 49-50, 52-53
    sampling properties and confidence intervals, 61-62
Partial area under the curve (PAUC), see also Area under the curve (AUC)
    curves comparison, 114
    hypothesis testing, 82
    prostate-specific antigen, 102
    standardizing, 29
    summary indices, 28-29, 69-70
    summary statistics adjustment, 99-100
Partially paired data, 122
Pattern recognition, 4, 35
PAUC, see Partial area under the curve (PAUC)
Permutation procedures, 119-120

Personality Assessment Inventory (PAI), 200
Personal loan data, see Credit scoring; Finance
Pierce skill scores, 192
Pima Indians
    empirical estimation, 45
    loss difference plots, 72
    nonparametric estimation, 54, 63
    parametric curve fitting, 50
Placement value, 68, 74
Pooling, covariates, 88
Populations
    atmospheric sciences, 184
    AUC, 26-28, 66-68, 77-78, 99
    Bayesian methods, 125, 134, 136-137
    bias, 165
    binary regression methods, 56
    binormal case, 115
    binormal model, 31-32, 34-35
    classification, 4-5
    classifier performance assessment, 7-8
    comparing curves, 111-114
    confidence intervals, 67-68, 79
    continuous scores, 23
    cost-weighted misclassification rates, 24
    covariates, 87
    definition, 18-19
    direct covariate adjustment, 93-95
    estimation, 41-43, 45, 59-60, 66
    fundamentals, 7-8, 17-18, 76
    further reading, 35
    general features, 19-20
    incremental value, 103
    indirect covariate adjustment, 89
    interpretation, 26-27
    Kolmogorov-Smirnov, empirical ROC, 76-77

maximum vertical distance, 30
measurement errors, 82-85
multiple classes, 150-151
Neyman-Pearson Lemma, 24-25
nonparametric estimation, 54
parametric curve fitting, 50
partial AUC, 69-70
properties, 20-23
single points and partial areas, 28-29
slope and optimality results, 24-25
summary indices, 25-30
uncertain/unknown group labels, 132-133
Youden Index, 30
Positive predictive value (ppv), 9
Posterior density, 126-127
Posterior distributions, 123-124, 127, 131
Posterior predictive density, 126
P population, see Populations
ppv, see Positive predictive value (ppv)
Precipitation, 186
Precision, 10, 142-143
Prediction model (diagnostic), 164
Prevalence, 9
Prior distributions, 123, 131
Probabilities (general), 18, see also specific type
Probit link function and transformation
  AUC adjustment, 98-99
  covariate adjustment, 96
  partial AUC adjustment, 100
Prognostic medicine, 6, see also Medical context and applications
Proportional hazards model, 175, 197
Prostate-specific antigen (PSA), 102
PSA, see Prostate-specific antigen (PSA)

Psychology, 74
Pulmonary embolism, 164

Quadratic discriminators and discriminant function
  classifier type, 39
  curves that cross, 107-108
  empirical estimation, 45, 47
  loss difference plots, 72
  nonparametric estimation, 54
Quasi-likelihood methods, 90, 91, 96

Radiography, 14
Radiology, 179
Random-effects framework, 128-129
Rare diseases, 10-11
Rate, 18, 28-29
Rating scale, 65-66
Recall, 8-10, 142-143
Reconviction, 200
References, 205-228, see also Further reading
Regression approaches
  AUC adjustment, 98-99
  comparing curves, 110
  design and interpretation issues, 179
  direct covariate adjustment, 92, 94
  entire curves comparison, 120-121
  incremental value, 103-104
  indirect covariate adjustment, 90-91
  parametric curve fitting, 62
  partial AUC adjustment, 99-100
Reject inference, 171-172
Relative operating level (ROL), 185-187
REML, see Restricted maximum likelihood (REML)
Representative objects, 5-6

INDEX

Restricted maximum likelihood (REML), 129-130
Restricted placement values, 74
Retinal patterns, 4
Revelation effect, 198
Reversing an axis, 141-142
Risk score, 103, 105
Rules, classification, 5

Sample size calculations, 78-82
Sampling distribution, 57-58
Sampling properties and confidence intervals
    binary regression methods, 63
    empirical estimator, 58-60
    fundamentals, 57-58
    nonparametric estimation, 62-63
    parametric curve fitting, 61-62
Scores
    AUC adjustment, 98
    Bayesian methods, 125-126
    continuous, 23
    incremental value, 103
    Neyman-Pearson Lemma, 25
    nonparametric approach, 118-119
Selection bias, 166-167, see also Bias
Semiparametric distribution-free approach, 53
Semiparametric efficient (SPE), 169
Semiparametric transformation models, 47
Sensitivity, 9
Separation tests, 76-78
Sequence randomization, 177
Sexual offender groups, reconviction, 200
Signal detection theory, 35
Single continuum, objects on a, 156
Single point rates, 28-29
Skewness, 54
Skill plot, 145
Slope
    binormal ROC curve, 32
    cost-weighted misclassification rates, 24
    Neyman-Pearson Lemma, 24-25
    optimal classification threshold, 70-71
    ROC properties, 22-23
Smoothing, nonparametric estimation, 56
Sociology applications, 199-200
Software, 201-202
SPE, see Semiparametric efficient (SPE)
Specificity, 9
Spectrum bias, 166, see also Bias
Speech recognition, 1, 4
Standard error, 58
Standard normal distribution, 32
Standard object classes, 5-6
Statistical inference, 38, see also Inference on single curves
Statistics, classification, 4
Stochastic imputation, 162
Structured Interview of Reported Symptoms (SIRS) scale, 200
Studies, bias in
    fundamentals, 165-166
    reject inference, 171-172
    verification bias, 166-171
Subintervals, AUC estimation, 65
Summary indices
    AUC, 26-28, 65-69
    confidence intervals, 67-69
    estimation, 65-67
    fundamentals, 25-26, 63
    LC index, 71-72, 74
    loss difference plots, 71-72, 74
    maximum vertical distance, 30
    optimal classification threshold, 70-71
    partial AUC, 69-70
    single points and partial areas, 28-29
    Youden Index, 30
Summary statistics and adjustment

AUC adjustment, 97-99
  curves comparison, 109-112
  fundamentals, 97
  partial AUC adjustment, 99-100
  Youden Index, 100-101
Supervised classification, 5
Support vector machines, 39
Survival analysis, 196-197
Survivor function, 197
Systolic blood pressure, 83

Taylor series expansion, 84
Ternary diagrams, 151
Tests
  AUC=0.5, 77-78
  blinding, medical imaging, 177
  of equality, 122
  fundamentals, 76
  Kolmogorov-Smirnov, empirical ROC, 76-77
  sets, 7, 40-41
  two-sample, 164-165
Time, variants to ROC curves, 155
Training data and sets, 5, 6, 6-7, 40-41
Transformation, 22, see also specific type
Transformed index, 29
Transformed normal method, 174
Trapezium rule, 65
Trapezoidal method, 164, 167
True positive/negative rates, 9-11
Tuning, 41
Two-by-two classification table, 8
Two-sample tests, 164-165
Two-sample transformation model, 53

UCI databases, 45, 183
Unbalanced situations, 143, 193
Unbiased properties, see Bias
Uncertain/unknown group labels
  fundamentals, 132-134
  nonparametric estimation, 137-139
  parametric estimation, 134-136
Unconditional imputation, 162
Univariate variables, 3
University classes application, 1
University of Chicago (Kurt Rossman Laboratories), 201
University of Iowa (Department of Radiology, Medical Image Perception Laboratory), 201
Unpaired data, 108, 118
Unrelated revelation, 198
U.S. National Weather Service, 189

Variability, empirical estimator, 59-60
Variables, classification, 2-3
Variances, 108, 129
Variants to ROC curves
  alternatives, 141-145
  convex hull, 145-147
  fundamentals, 141
  issues, 154-155
  multiple classes, 147-154
  time, 155
Verification bias, see also Bias
  bias in studies, 166-171
  defined, 158
  missing values, 163
Vineland Adaptive Behavior Scales, 200
Violent offender groups, reconviction, 200
Volume under the surface (VUS), 151, 156
VUS, see Volume under the surface (VUS)

Wald statistic, 178-179
Wealth, 27-28
Weather, see Atmospheric sciences
Weibull model and distribution
  indirect covariate adjustment, 90

INDEX

  parametric curve fitting, 53
  survival analysis, 197
Weight function, 114
Wilcoxin rank-sum statistic, 170
WinBUGS software, 124, 135
Within-study variances, 129
Work-up bias, 158, 167, see also
    Bias

Xing filtering, 195

Youden Index
  classifier performance assessment,
    11
  fundamentals, 30, 100-101
  Kolmogorov-Smirnov test, 77
  optimum threshold choice, 173
  summary indices, 30
  summary statistics adjustment,
    100-101